Manual of Water Supply Practices

M45

Fiberglass Pipe Design

AF895549

Fourth Edition

American Water Works Association

Manual of Water Supply Practices—M45, Fourth Edition

Fiberglass Pipe Design

Disclaimer

The authors, contributors, editors, and publisher do not assume responsibility for the validity of the content or any consequences of their use. In no event will AWWA be liable for direct, indirect, special, incidental, or consequential damages arising out of the use of information presented in this book. In particular, AWWA will not be responsible for any costs, including, but not limited to, those incurred as a result of lost revenue. In no event shall AWWA's liability exceed the amount paid for the purchase of this book.

Sr. Editorial Manager, Books: Suzanne Snyder
Manager – Manuals: Molly Beach
Manager – Publishing Operations: Gillian Wink
Specialist – Copyright and Permissions: Peggy Tyler
Director of Publishing: John Fedor
Technical Editor: Wiley
Cover Image: Mark Grenier/Shutterstock.com
Production: Innodata

Library of Congress Cataloging-in-Publication Data

Names: American Water Works Association, editor.
Title: M45 - Fiberglass Pipe Design / [edited] by American Water Works Association.
Description: First edition. | Denver, CO : American Water Works Association, [2026] | Includes bibliographical references and index. |
Identifiers: LCCN 2024030791 (print) | LCCN 2024030792 (ebook) | ISBN 9781647171957 (paperback) | ISBN 9781613007204 (pdf)
Subjects: LCSH: Water--Purification--Biological treatment.
Classification: LCC TD475 .M28 2025 (print) | LCC TD475 (ebook) | DDC 628.1/62--dc23/eng/20241023
LC record available at https://lccn.loc.gov/2024030791
LC ebook record available at https://lccn.loc.gov/2024030792

ISBN 978-1-64717-200-8

ISBN, electronic 978-1-61300-725-9

http://doi.org/10.12999/AWWA.M45ed4

American Water Works Association

American Water Works Association
6666 West Quincy Avenue
Denver, CO 80235-3098
awwa.org

Contents

Figures

Tables

This page intentionally blank.

Preface

This manual is the fourth edition of AWWA Manual M45, *Fiberglass Pipe Design*. It provides the reader with both technical and general information to aid in the design, specification, procurement, installation, and understanding of fiberglass pipe and fittings. The manual is a discussion of recommended practice, not an AWWA standard calling for compliance with certain requirements. It is intended for use by utilities and municipalities of all sizes, whether as a reference book or textbook for those not fully familiar with fiberglass pipe and fitting products. Design engineers and consultants may use this manual in preparing plans and specifications for new fiberglass pipe design projects.

The manual covers fiberglass pipe and fitting products and certain appurtenances and their application to practical installations, whether of a standard or special nature. For adequate knowledge of these products, the entire manual should be studied. Readers will also find the manual a useful source of information when assistance is needed with specific or unusual conditions. The manual contains a list of applicable national standards, which may be purchased from the respective standards organizations (e.g., AWWA, ASTM, etc.).

This fourth edition includes updates to trench and soil terminology plus compaction definition changes to achieve uniformity among all piping materials. Several typographical errors were corrected.

This page intentionally blank.

Acknowledgments

The American Water Works Association (AWWA) *Fiberglass Pipe Design* Manual Subcommittee, which developed this edition of the M45 manual, had the following personnel at the time:

Richard C. Turkopp, Chair

A. Dail (liaison, nonvoting), Staff Engineer Liaison, AWWA, Denver, Colo.

R.A. Johnson, Russcor Engineering, Naples, Fla.

N.E. Kampbell, Rehabilitation Resource Solutions LLC, Hilliard, Ohio

D.P. Kozman, Hammerhead Trenchless Equipment, Lake Mills, Wis.

J.A. LeBlanc, Thompson Pipe Group, Baton Rouge, La.

A. Marner, US Bureau of Reclamation, Denver, Colo.

R.N. Renkes, Fiberglass Tank and Pipe Institute, Tulsa, Okla.

P.A. Sharff, Simpson Gumpertz and Heger, Inc., Waltham, Mass.

R.C. Turkopp, Kingwood, Tex.

This edition of the manual was also reviewed and approved by the AWWA Standards Council and the Standards Committee on Thermosetting Fiberglass Reinforced Plastic Pipe. The Standards Committee on Thermosetting Fiberglass Reinforced Plastic Pipe had the following personnel at the time of approval:

Phillip A. Sharff, Chair
Matthew C. Richie, Vice-Chair

Utility Members

N. Gan, Region of Peel Public Works, Brampton, Ont.

A. Marner, US Bureau of Reclamation, Denver, Colo.

General Interest Members

M.M. Baig, Saudi Aramco, Dhahran, Saudi Arabia

A. Chastain-Howley (liaison, nonvoting), Black & Veatch, Camden, Maine

A. Dail (liaison, nonvoting), Staff Engineer Liaison, AWWA, Denver, Colo.

R.A. Johnson, Russcor Engineering, Naples, Fla.

N.E. Kampbell, Rehabilitation Resource Solutions LLC, Hilliard, Ohio

P.A. Sharff, Simpson Gumpertz and Heger, Inc., Waltham, Mass.

Producer Members

R.N. Renkes, Fiberglass Tank and Pipe Institute, Tulsa, Okla.

R.C. Turkopp, Kingwood, Tex.

This page intentionally blank.

Chapter 1

History and Use

1.1 INTRODUCTION

Fiberglass pipe is made from glass fiber reinforcements embedded in or surrounded by cured thermosetting resin. This composite structure may also contain aggregate, granular, or platelet fillers; thixotropic agents; and pigments or dyes. By selecting the proper combination of resin, glass fibers, fillers, and design, the fabricator can create a product that offers a broad range of properties and performance characteristics. Over the years, the diversity and versatility of materials used to manufacture fiberglass pipe have led to a variety of names for fiberglass pipe. Among these are reinforced thermosetting-resin pipe, reinforced polymer mortar pipe, fiberglass-reinforced epoxy, glass-reinforced plastic, and fiberglass-reinforced plastic. Fiberglass pipes have also been categorized by the particular manufacturing process—filament winding or centrifugal casting. Frequently, the resin used to manufacture the fiberglass pipe—epoxy, polyester, or vinyl ester—has been used to classify or grade fiberglass pipes.

Regardless of the many possible combinations, the most common and useful designation is simply "fiberglass pipe." This name encompasses all the various available products and allows consideration as a unique and general class of engineering materials.

1.2 HISTORY

Fiberglass pipe was introduced in 1948. The earliest application for fiberglass piping and still one of the most widely used is in the oil industry. Fiberglass pipe was selected as a corrosion-resistant alternative to protected steel, stainless steel, and other more exotic metals. Product lines expanded to include applications of increasingly high pressure and downhole tubing with threaded connections. In the late 1950s, larger diameters became available, and fiberglass pipe was increasingly used in the chemical process industry because of the pipe's inherent corrosion-resistant characteristics.

Since the 1960s, fiberglass pipe products have been used for municipal water and sewage applications. Fiberglass pipe combines the benefits of durability, strength, and corrosion resistance, thus eliminating the need for interior linings, exterior coatings, and cathodic protection. Fiberglass pipe systems offer great design flexibility with a wide range of standard pipe diameters and fittings available, as well as an inherent ability for custom fabrication to meet special needs. Fiberglass pipe is available in diameters ranging from 1 through 156 in. (25–4,000 mm). Fiberglass pipe is available in pressure classes ranging from gravity applications through several thousand lbs per square in. (kilopascals). Few countries in the world have not used fiberglass pipe.

1.3 APPLICATIONS

Fiberglass pipe is used in many industries and for a myriad of applications, including

- chemical processes,
- desalination,
- downhole tubing and casing,
- ducting and vent piping,
- geothermal piping,
- industrial effluents,
- irrigation,
- oil fields,
- potable water,
- power plant cooling and raw water,
- sanitary sewers,
- seawater intake and outfalls,
- slurry piping,
- storm sewers,
- water distribution, and
- water transmission.

1.4 STANDARDS, SPECIFICATIONS, AND REFERENCE DOCUMENTS

Many organizations have published nationally recognized standards, test methods, specifications, and recommended practices on fiberglass piping systems and products. These organizations include ASTM International (ASTM), the American Petroleum Institute (API), the American Society of Mechanical Engineers (ASME), NSF International (NSF), UL Standards & Engagement (ULSE), Factory Mutual (FM), the American National Standards Institute (ANSI), and the International Organization for Standardization (ISO).

Following is a list of fiberglass pipe standards and specifications that are commonly used in specifying, testing, and using fiberglass piping systems.

1.4.1 Product Specifications and Classifications

General.

ASTM D2517	Standard Specification for Reinforced Epoxy Resin Gas Pressure Pipe and Fittings

ASTM D2996 Standard Specification for Filament-Wound "Fiberglass" (Glass-Fiber-Reinforced Thermosetting-Resin) Pipe (Applicable to epoxy, polyester, and furan resins in sizes from 1 to 16 in. [25 to 400 mm])

ASTM D2997 Standard Specification for Centrifugally Cast "Fiberglass" (Glass-Fiber-Reinforced Thermosetting-Resin) Pipe (Applicable for 1 through 14 in. [25 through 350 mm] pipe of polyester or epoxy resins)

ASTM D3262 Standard Specification for "Fiberglass" (Glass-Fiber-Reinforced Thermosetting-Resin) Sewer Pipe (Applicable for pipes 8 through 156 in. [200 through 4,000 mm] diameter, with or without siliceous sand, and polyester or epoxy resin)

ASTM D3517 Standard Specification for "Fiberglass" (Glass-Fiber-Reinforced Thermosetting-Resin) Pressure Pipe (Applicable for pipes 8 through 156 in. [200 through 4,000 mm] diameter, with or without siliceous sand, and polyester or epoxy resin)

ASTM D3754 Standard Specification for "Fiberglass" (Glass-Fiber-Reinforced Thermosetting-Resin) Sewer and Industrial Pressure Pipe (Applicable for 8 through 156 in. [200 through 4,000 mm] diameter, with or without siliceous sand, and polyester or epoxy resin)

ASTM D4024 Standard Specification for Machine Made "Fiberglass" (Glass-Fiber-Reinforced Thermosetting Resin) Flanges (Applicable for ½ through 24 in. [13 through 600 mm] ANSI B16.5 150 lb [70 kg] bolt circle flanges)

ASTM D4161 Standard Specification for "Fiberglass" (Glass-Fiber-Reinforced Thermosetting-Resin) Pipe Joints Using Flexible Elastomeric Seals

ASTM F1173 Standard Specification for Thermosetting Resin Fiberglass Pipe Systems to Be Used for Marine Applications

API 15LR Specification for Low Pressure Fiberglass Line Pipe (Applicable to 2 through 24 in. [50 through 600 mm] diameter pipe of epoxy or polyester resin for use at cyclic pressures to 1,000 psi [6,895 kPa])

API 15HR Specification for High Pressure Fiberglass Line Pipe (Applicable to 1 through 10 in. [25 through 250 mm] pipe and fittings for operating pressures of 500 psi [3,500 kPa] to 5,000 psi [35,000 kPa])

ANSI/AWWA C950 AWWA Standard for Fiberglass Pressure Pipe

US military (MIL) specifications.

MIL-P-24608 Pipe, Fittings, and Adhesive Kits, Glass Reinforced Thermosetting Epoxy Resin for Shipboard Piping Systems (Specification for epoxy resin pipe from ½ through 12 in. [13 through 300 mm] diameters for 200 psig [1,379 kPa] service at 150°F [66°C] for US Navy shipboard applications)

MIL-P-28584A Pipe and Pipe Fittings, Glass Fiber Reinforced Plastic, Adhesive Bonded Joint Type, for Condensate Return Lines (Specification for epoxy resin pipe and fittings from 2 through 12 in. (50 through 300 mm) diameter for use as steam condensate return lines in continuous service at 125 psig [862 kPa] and 250°F [121°C])

MIL-P-29206A Pipe and Pipe Fittings, Glass Fiber Reinforced Plastic, for Liquid Petroleum Lines (Specification for epoxy or polyester pipe and fittings 2 through 12 in. [50 through 300 mm] in diameter for POL services to 150°F [66°C] and 150 psig (1,034 kPa) with surges to 250 psig [1,724 kPa])

1.4.2 Recommended Practices

Dimensions.

ASTM D3567	Standard Practice for Determining Dimensions of "Fiberglass" (Glass-Fiber-Reinforced Thermosetting-Resin) Pipe and Fittings

Installation.

ASTM D3839	Standard Guide for Underground Installation of "Fiberglass" (Glass-Fiber-Reinforced Thermosetting-Resin) Pipe
API RP 15TL4	Recommended Practice for Care and Use of Fiberglass Tubulars
API RP 1615	Installation of Underground Hazardous Substances or Petroleum Storage Systems

1.4.3 Standard Test Methods

Tensile properties.

ASTM D638	Standard Test Method for Tensile Properties of Plastics
ASTM D1599	Standard Test Method for Resistance to Short-Time Hydraulic Failure Pressure of Plastic Pipe, Tubing, and Fittings
ASTM D2105	Standard Test Method for Longitudinal Tensile Properties of Fiberglass (Glass-Fiber-Reinforced Thermosetting-Resin) Pipe and Tube
ASTM D2290	Standard Test Method for Apparent Hoop Tensile Strength of Plastic or Reinforced Plastic Pipe

Compressive properties.

ASTM D695	Standard Test Method for Compressive Properties of Rigid Plastics

Bending properties.

ASTM D790	Standard Test Methods for Flexural Properties of Unreinforced and Reinforced Plastics and Electrical Insulating Materials
ASTM D2925	Standard Test Method for Beam Deflection of "Fiberglass" (Glass-Fiber-Reinforced Thermosetting Resin) Pipe Under Full Bore Flow

Long-term internal pressure strength.

ASTM D1598	Standard Test Method for Time-to-Failure of Plastic Pipe Under Constant Internal Pressure
ASTM D2143	Standard Test Method for Cyclic Pressure Strength of Reinforced, Thermosetting Plastic Pipe
ASTM D2992	Standard Practice for Obtaining Hydrostatic or Pressure Design Basis for "Fiberglass" (Glass-Fiber-Reinforced Thermosetting-Resin) Pipe and Fittings

Pipe stiffness.

ASTM D2412	Standard Test Method for Determination of External Loading Characteristics of Plastic Pipe by Parallel-Plate Loading

External pressure.

ASTM D2924	Standard Test Method for External Pressure Resistance of "Fiberglass" (Glass-Fiber-Reinforced Thermosetting-Resin) Pipe

Chemical resistance.

ASTM C581	Standard Practice for Determining Chemical Resistance of Thermosetting Resins Used in Glass-Fiber-Reinforced Structures Intended for Liquid Service
ASTM D3681	Standard Test Method for Chemical Resistance of "Fiberglass" (Glass–Fiber–Reinforced Thermosetting-Resin) Pipe in a Deflected Condition
ASTM D5365	Standard Test Method for Long-Term Ring-Bending Strain of "Fiberglass" (Glass-Fiber-Reinforced Thermosetting-Resin) Pipe

1.4.4 Product Listings, Approvals, and Piping Codes

NSF International—Standard Numbers 14 and 61.	Tests and lists fiberglass pipe, fittings, and adhesives for use in conveying potable water; additionally, tests and certifies products as to their classification to an applicable national standard or for special properties (Standard 14 only).
UL Solutions.	Provides testing, inspection, certification, and listing for fiberglass pipe.
UL Standards & Engagement Inc.	Provides established standards for fiberglass pipe for use as underground fire water mains and underground transport of petroleum products.
Factory Mutual Research.	Has established an approval standard for plastic pipe and fittings for underground fire protection service.
ANSI/ASME B31.1— Power Piping.	This code prescribes minimum requirements for the design, materials, fabrication, erection, testing, and inspection of power and auxiliary service piping systems for electric generation stations, industrial institutional plants, and central and district heating plants.
ANSI/ASME B31.3— Process Piping.	This code lists some ASTM, AWWA, and API fiberglass pipe specifications as acceptable for use within the code and establishes criteria for their installation and use. This code, in addition to other ASME codes, establishes rules regarding the application of fiberglass piping and provides engineering guidance for the use of fiberglass materials.
ANSI/ASME B31.8— Gas Transmission and Distribution Piping Systems.	This code lists fiberglass pipe manufactured in compliance with ASTM D2517 as acceptable for use within the code.
Department of Transportation, Title 49, Part 192.	This is a code of federal regulations that covers the transportation of natural and other gases by pipeline. Minimum federal standards are included.
ASME Boiler and Pressure Vessel Code Case N155.	This code provides the rules for the construction of fiberglass piping systems for use in Section III, Division I, Class 3 applications in nuclear power plants.

1.4.5 ISO Standards and Specifications

ISO has issued many standards, test methods, and technical reports relating to fiberglass piping systems and products. Many of their titles, as well as the general content, are very similar to the US-issued standards covered previously.

Product specifications.

ISO 10467	Plastics piping systems for pressure and nonpressure drainage and sewerage—Glass-reinforced thermosetting plastics (GRP) systems based on unsaturated polyester (UP) resin
ISO 10639	Plastics piping systems for pressure and nonpressure water supply—Glass-reinforced thermosetting plastics (GRP) systems based on unsaturated polyester (UP) resin
ISO 14692	Petroleum and natural gas industries—Glass-reinforced plastics (GRP) piping

Test methods.

ISO 7432	Glass-reinforced thermosetting plastics (GRP) pipes and fittings—Test methods to prove the design of locked socket-and-spigot joints, including double-socket joints, with elastomeric seals
ISO 7509	Plastics piping systems—Glass-reinforced thermosetting plastics (GRP) pipes—Determination of time to failure under sustained internal pressure
ISO 7510	Plastics piping systems—Glass-reinforced plastics (GRP) components—Determination of the amounts of constituents
ISO 7511	Plastics piping systems—Glass-reinforced thermosetting plastics (GRP) pipes and fittings—Test methods to prove the leaktightness of the wall under short-term internal pressure
ISO 7684	Plastics piping systems—Glass-reinforced thermosetting plastics (GRP) pipes—Determination of the creep factor under dry conditions
ISO 7685	Glass-reinforced thermosetting plastics (GRP) pipes—Determination of initial specific ring stiffness
ISO 8483	Glass-reinforced thermosetting plastics (GRP) pipes and fittings—Test methods to prove the design of bolted flange joints
ISO 8513	Plastics piping systems—Glass-reinforced thermosetting plastics (GRP) pipes—Test methods for the determination of the initial longitudinal tensile strength
ISO 8521	Glass-reinforced thermosetting plastics (GRP) pipes—Test methods for the determination of the initial circumferential tensile wall strength
ISO 8533	Glass-reinforced thermosetting plastics (GRP) pipes and fittings—Test methods to prove the design of cemented or wrapped joints
ISO 8639	Glass-reinforced thermosetting plastics (GRP) pipes and fittings—Test methods for leaktightness and proof of structural design of flexible joints
ISO 10466	Glass-reinforced thermosetting plastics (GRP) pipes—Test method to prove the resistance to initial ring deflection
ISO 10468	Glass-reinforced thermosetting plastics (GRP) pipes—Determination of the ring creep properties under wet or dry conditions

ISO 10471	Glass-reinforced thermosetting plastics (GRP) pipes—Determination of the long-term ultimate bending strain and the long-term ultimate relative ring deflection under wet conditions
ISO 10928	Plastics piping systems—Glass-reinforced thermosetting plastics (GRP) pipes and fittings—Methods for regression analysis and their use
ISO 10952	Glass-reinforced thermosetting plastics (GRP) pipes and fittings—Determination of the resistance to chemical attack for the inside of a section in a deflected condition
ISO 14828	Glass-reinforced thermosetting plastics (GRP) pipes—Determination of the long-term specific ring relaxation stiffness under wet conditions and calculation of the wet relaxation factor
ISO 15306	Glass-reinforced thermosetting plastics (GRP) pipes—Determination of the resistance to cyclic internal pressure

Technical reports.

ISO/TR 10465-1	Underground installation of flexible glass-reinforced thermosetting resin (GRP) pipes—Part 1: Installation procedures
ISO/TR 10465-2	Underground installation of flexible glass-reinforced thermosetting resin (GRP) pipes—Part 2: Comparison of static calculation methods
ISO/TR 10465-3	Underground installation of flexible glass-reinforced thermosetting resin (GRP) pipes—Part 3: Installation parameters and application limits

1.5 TERMINOLOGY

Fiberglass pipe users may encounter some unique or unfamiliar terminology. A glossary of terms used in this manual and by those in the fiberglass pipe industry is provided at the end of this manual.

Following is a list of additional references for terminology commonly used in the fiberglass pipe industry.

1.5.1 Terminology Reference Documents

ASTM D8	Standard Terminology Relating to Materials for Roads and Pavements
ASTM D653	Standard Terminology Relating to Soil, Rock, and Contained Fluids
ASTM D883	Standard Terminology Relating to Plastics
ASTM F412	Standard Terminology Relating to Plastic Piping Systems

This page intentionally blank.

Chapter 2

Materials, Properties, and Characteristics

2.1 GENERAL

Fiberglass pipe is a composite material system produced from glass fiber reinforcements, thermosetting plastic resins, and additives. By selecting the right combination and amount of materials and the specific manufacturing process, the designer can create a product to meet the most demanding requirements. The result is a material with a broad range of characteristics and performance attributes.

2.2 CHARACTERISTICS

The following is a list of general characteristics of fiberglass composite pipe.

Corrosion resistance. Fiberglass pipe systems are resistant to corrosion, both inside and out, in a wide range of fluid-handling applications. As a result, additional linings and exterior coatings are not required.

Strength-to-weight ratio. Fiberglass composite piping systems have excellent strength-to-weight properties. The ratio of strength per unit of weight of fiberglass composites is greater than that of iron, carbon, and stainless steels.

Lightweight. Fiberglass composites are lightweight. Fiberglass piping is approximately one-sixth the weight of similar steel products and one-tenth the weight of similar concrete products.

Electrical properties. Standard fiberglass pipes are nonconductive. Some manufacturers offer conductive fiberglass piping systems for applications that require dissipation of static electricity buildup when transporting certain fluids like jet fuel.

Dimensional stability. Fiberglass composites can maintain the critical tolerances required of the most demanding structural and piping applications. The material meets the most stringent material stiffness, dimensional tolerance, weight, and cost criteria.

Low maintenance cost. Fiberglass piping is easy to maintain because it does not rust, is easily cleaned, and requires minimal protection from the environment.

2.3 THE MATERIAL SYSTEM

Fiberglass composites consist of glass fiber reinforcements, thermosetting resins, and additives, which are designed and processed to meet specific functional performance criteria. To aid understanding of the performance characteristics of a finished fiberglass pipe, the interrelationship of the system components is outlined in this chapter. The following is a list of terms used in describing the material system.

Fiberglass reinforcement. This term indicates the amount, type, location, and orientation of glass fibers in the pipe that will provide the required mechanical strength.

Resin system. Resin selection will provide the physical and chemical properties (e.g., the glass transition temperature, a measurement of resistance to heat, and softening or plasticization by solvents and gases).

Following is a brief review of the constituents of fiberglass pipe and how they influence the finished pipe product.

2.3.1 Glass Fiber Reinforcements

The mechanical strength of fiberglass pipe depends on the amount, type, and arrangement of glass fiber reinforcement. Strength increases proportionally with the amount of glass fiber reinforcement. The quantity of the glass fibers (and the direction in which the individual strands are placed) determines the strength.

2.3.1.1 Fiberglass types. Fiberglass materials are available with a variety of compositions. This allows for additional design flexibility to meet performance criteria. All fiberglass reinforcement begins as individual filaments of glass drawn from a furnace of molten glass. Many filaments are formed simultaneously and gathered into a "strand." A surface treatment (sizing) is added to maintain fiber integrity, establish compatibility with resin, and ease further processing by improving consolidation and wet strength. Sizing can also affect resin chemistry and laminate properties.

The glass fibers most commonly used in pipe are referred to as Types E, ECR, and C. Glass types ECR and C provide improved acid and chemical resistance. Type C glass fibers are generally only used to reinforce chemical-resistant liners.

2.3.1.2 Fiberglass reinforcement forms. Following is a brief description of the various forms of fiberglass reinforcements.

Continuous roving. This consists of bundled, untwisted strands of glass fiber reinforcement and comes as cylindrical packages for further processing. Continuous roving is typically used in filament winding and unidirectional/bidirectional reinforcements and may be processed into chopped strand mat used to provide multidirectional reinforcement in pipe and fittings.

Woven roving. This is a heavy, drapable fabric woven from continuous roving. It is available in various widths, thicknesses, and weights. Woven roving provides high strength to large molded parts and is lower in cost than conventional woven fabrics.

Reinforcing mats. These are chopped strands held together with resinous binders. Two kinds of reinforcing mats are used in pipe and fittings (i.e., chopped strand mat and woven roving combination mat). Chopped strand mats are used in medium-strength applications for pipe fittings and reinforcing, where a uniform cross section is desired. Use of the combination mat saves time in hand lay-up operations.

Surface veils. These lightweight fiberglass reinforcement mats allow layers with a high resin content with minimal reinforcement. The surface veil provides extra environmental resistance for pipe and fittings, plus a smooth appearance. (Some surface veils from polyester fibers are also used.)

2.3.1.3 Reinforcement arrangement. The three general types of fiber orientation include the following:

Unidirectional. The greatest strength is in the direction of the fibers. Up to 80 percent reinforcement content by weight is possible.

Bidirectional. Some fibers are positioned at an angle to the rest of the fibers, as with helical filament winding and woven fabrics. This provides different strength levels governed by the fiber quantity in each direction of fiber orientation. A combination of continuous and chopped fibers is also used to provide designed directional strength.

Multidirectional (isotropic). This arrangement provides nearly equal, although generally lower, strength and modulus in all directions. Ten to 50 percent reinforcement content by weight can be obtained with multidirectional materials such as chopped roving or chopped strand mat.

2.3.2 Resins

The second major component of fiberglass pipe is the resin system. Manufacturers choose a resin system for chemical, mechanical, and thermal properties and processability.

The two basic groups of resin systems are thermosetting and thermoplastic. Fiberglass pipe, by definition, uses only thermosetting-resin systems. Thermosets are polymeric resin systems cured by heat or chemical additives. Once cured, a thermoset is essentially infusible (cannot be remelted) and insoluble.

The thermosetting resins used in fiberglass pipe fall into two general categories—polyesters and epoxies.

2.3.2.1 Polyester resins. Polyester resins are commonly used to produce large-diameter water and sewage piping. Polyesters have excellent water and chemical resistance and are noted for acid resistance.

The base polyester resin is a solid. It is typically dissolved in styrene monomer, with which it cross-links to provide the final thermoset structure. Polyester resins are cured by organic peroxide catalysts. The type and amount of catalyst will influence gel time, cure time, curing temperature, and the degree of cure. Typical catalysts include methyl ethyl ketone peroxide and benzyl peroxide. Manufacturers may select from several different types of polyester resins that provide a wide range of performance characteristics. These include

- orthophthalic polyester,
- terephthalic polyester,
- chlorendic acid polyester,
- novolac epoxy vinyl ester,
- isophthalic polyester,
- bisphenol-A fumarate polyester, and
- bisphenol-A vinyl ester.

2.3.2.2 Epoxy resins. Epoxy resins are commonly used in the manufacture of smaller-diameter piping (<30 in. [800 mm]) conveying water, condensates, hydrocarbons, caustics, and dilute acids. Fiberglass epoxy piping is used in oil fields at pressures up to several thousand pounds per square inch (kilopascals). Epoxy resins typically allow higher service temperatures than polyester resins, ranging up to about 225°F (108°C).

Epoxy resins cannot be categorized by resin type as easily as polyesters. The type of curing agent, or hardener, is critical with epoxy resins because the agent influences the composite properties and performance. The two basic types are amine- and anhydride-cured bisphenol-A epoxies.

Bisphenol-A epoxy resins are commonly cured with multifunctional primary amines. For optimum chemical resistance, these mixtures usually require a heat cure and/or post cure. The cured resin has good chemical resistance, particularly in alkaline environments, and can have good temperature resistance. Bisphenol-A epoxy resins may also be cross-linked with various anhydrides by using a tertiary amine accelerator and heat. These cured polymers generally have good chemical resistance, especially to acids.

2.4 OTHER COMPONENTS

Glass fiber reinforcements and thermosetting resins are the major constituents in fiberglass pipe. However, other materials that influence processing and/or product performance are used, including fillers, promoters, accelerators, inhibitors, and pigments.

Fillers. Inorganic materials, such as hydrated alumina, glass microspheres, clay, talc, calcium carbonate, sand, and calcium silicate, may yield economic, appearance, or performance advantages in fiberglass pipe.

Promoters, accelerators, and inhibitors. Promoters and accelerators advance the action of the catalyst to reduce the processing time. Inhibitors provide control over the cure cycle and increase the shelf life of the resin mix.

Pigments. The pigment choice affects the difference in reflected and transmitted color; clarity of the resin mix; reaction between dyes and other additives, such as catalysts; and the end-product color fastness and heat resistance.

2.5 PHYSICAL PROPERTIES

The following is a description of the physical properties of fiberglass pipe.

2.5.1 Chemical Resistance

All fiberglass pipes provide excellent resistance to water and native ground conditions. They are not subject to general corrosion attack, galvanic corrosion, aerobic corrosion, pitting, dezincification, and graphitic and intergranular corrosion. Fiberglass pipes are subject to some environmental stress and aging effects, the determination of which is part of the fiberglass pipe design procedure (see Chapter 5).

Fiberglass pipe resists a wide range of chemicals. The chemical resistance of fiberglass pipe depends primarily on the particular resin matrix material used. Although other factors, such as liner construction, cure, and fabrication method, may influence the chemical resistance of fiberglass pipe, the primary factor is the resin. The resins can be selected to provide chemical resistance to a broad range of materials. The fiberglass pipe manufacturer should be consulted for performance information for a particular chemical application.

2.5.2 Temperature Resistance

The temperature resistance of fiberglass pipe also depends largely on the resin matrix. The allowable upper limit of service temperature will also be influenced by the chemical environment and the stress condition of the piping system. In general, chemical agents are more aggressive at higher concentrations and elevated temperatures. However, for typical temperatures encountered in water supply systems (33–90°F [1–32°C]), fiberglass pipe is unaffected by service temperature, and the fiberglass pipe pressure performance does not need to be rerated or derated. Fiberglass pipe is virtually unaffected by colder temperatures. Therefore, normal shipping, handling, and storage procedures, as discussed in Chapter 10, may be used in subzero weather. However, users and installers of fiberglass pipe should be aware that the coefficient of thermal expansion for fiberglass pipe is generally higher than that for metal pipes (see Table 2-1). This must be recognized, and provisions must be made in design and installation to accommodate expansion and contraction, particularly in aboveground applications.

2.5.3 Abrasion Resistance

Fiberglass pipe provides generally good abrasion resistance and can be custom made for extremely abrasive service by lining the pipe with sand, silica flour, carborundum, or ceramic beads or tiles or by incorporating resilient liner materials such as polyurethanes. Special lining materials should match or exceed the hardness and abrasiveness of the contents being transported through the pipe or provide a high level of toughness and resilience.

2.5.4 Flame Retardants

The thermosetting-resin systems used to fabricate fiberglass pipes are organic materials. Therefore, under the proper combination of heat and oxygen, a thermosetting resin, like any organic matter, will burn. If required, the fire performance of fiberglass pipe can be enhanced by using resin systems that contain halogens or phosphorus. Use of hydrated fillers also enhances flame resistance. Other additives, primarily antimony oxides, can also increase the effectiveness of halogenated resins.

Fire performance testing requires small samples and specialized test methods and may not indicate how a material will perform in a full-scale field or fire situation. The fiberglass pipe manufacturer should be consulted for specific information on the combustion performance of fiberglass pipe.

Table 2-1 Mechanical properties range

	Units		Resin		Glass Fiber		Fiberglass Pipe	
Property	*in.-lb*	*SI*	*in.-lb*	*SI*	*in.-lb*	*SI*	*in.-lb*	*SI*
Tensile strength	10^3 psi	MPa	9.0–13.0	62–90	250–350	1,725–2,400	2.0–80.0	14–550
Tensile modulus	10^6 psi	GPa	0.4–0.6	2.8–4.1	10.0–11.0	69–75	0.5–5.0	3.5–34.5
Flexural strength	10^3 psi	MPa	10.0–22	69–150	250–350	1,725–2,400	4.0–70.0	28–480
Flexural modulus	10^6 psi	GPa	0.4–0.6	2.8–4.1	10.0–11.0	69–76	1.0–5.0	6.9–34.5
Coefficient of thermal expansion	in./in./°F $\times 10^{-6}$	mm/mm/°C $\times 10^{-6}$	11.0–55	20–100	3.0–3.3	5.4–6.0	8.0–30.0	14–54
Specific gravity	—	—	1.0–1.2	1.0–1.2	2.57–2.63	2.57–2.63	1.2–2.3	1.2–2.3
Compressive strength	10^3 psi	MPa	10.0–22	69–150	250–350	1,725–2,400	10.0–40.0	69–275

2.5.5 Weathering Resistance

Most thermosetting-resin systems used to fabricate fiberglass pipe are subject to some degradation from ultraviolet (UV) light. This degradation, however, is almost entirely a surface phenomenon. The structural integrity of fiberglass pipe is not affected by exposure to UV light. The use of pigments, dyes, fillers, or UV stabilizers in the resin system or painting of exposed surfaces can help reduce significantly any UV surface degradation. Surfaces exposed to UV light are generally fabricated with a resin-rich layer. Other weathering effects, such as rain or saltwater, are resisted fully by the inherent corrosion resistance of fiberglass pipe.

2.5.6 Resistance to Biological Attack

Fiberglass pipe will not deteriorate or break down under attack from bacteria or other microorganisms, nor will it serve as a nutrient to microorganisms, macroorganisms, or fungi. No known cases exist in which fiberglass pipe products suffered degradation or deterioration due to biological action. No special engineering or installation procedures are required to protect fiberglass pipe from biological attack. Some elastomers used in gaskets may be susceptible to this type of attack.

2.5.7 Tuberculation

Soluble encrustants, such as calcium carbonate, in some water supplies do not tend to precipitate onto the smooth walls of fiberglass pipe. Because fiberglass pipe is inherently corrosion resistant, tuberculation of the fiberglass pipe is not caused by corrosion byproducts.

2.6 MECHANICAL PROPERTIES

2.6.1 Mechanical Property Range

The design flexibility inherent with glass fiber–reinforced plastic materials and the range of manufacturing processes used preclude the simple listing of fiberglass pipe mechanical properties. For this reason, fiberglass pipe product standards are based on performance and detail product performance requirements rather than thickness property tables. Table 2-1 illustrates the broad range of mechanical properties available for resin, glass fiber, and fiberglass pipe.

This broad range of mechanical properties is further illustrated by the widely variable stress-strain curves possible with fiberglass pipe, depending on the amount, type, and orientation of the reinforcement as well as the manufacturing process. Figures 2-1 and 2-2 show the typical shape of the stress-strain curves for high- and low-pressure pipes for the circumferential and axial directions, respectively.

2.6.2 Mechanical Property Testing

The time dependence and wide range of mechanical properties of fiberglass pipe necessitate testing to develop data needed for design and analysis. Many test methods develop data over a moderate time range and then statistically extrapolate the data to establish long-term design values.

For example, the key long-term property test for fiberglass pipe is the development of a hydrostatic design basis (HDB) to establish the pipe pressure rating. This testing is conducted in accordance with ASTM D2992, Standard Practice for Obtaining Hydrostatic or Pressure Design Basis for "Fiberglass" (Glass-Fiber-Reinforced Thermosetting-Resin) Pipe and Fittings.

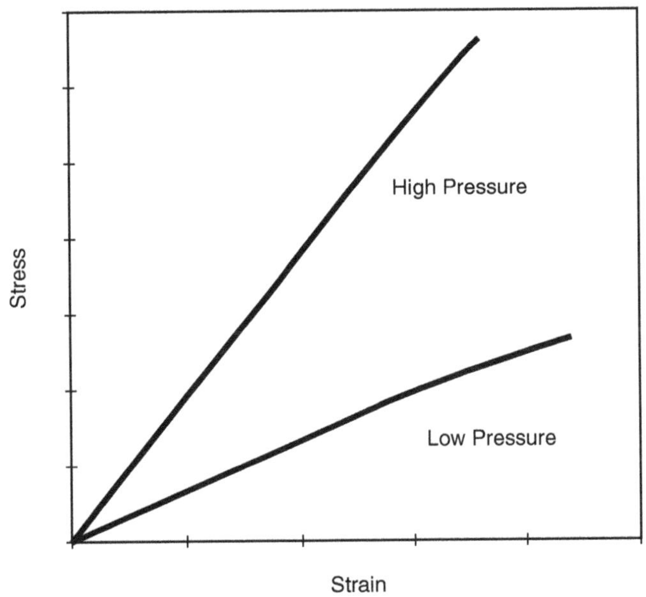

Figure 2-1 Typical circumferential stress–strain curves

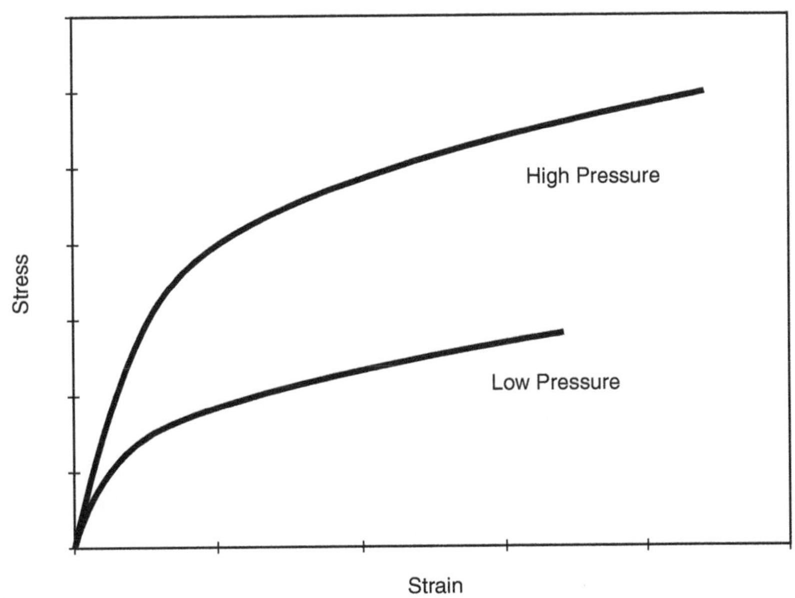

Figure 2-2 Typical axial stress–strain curves

This method requires pressurizing a minimum of 18 pipe samples at pressures far exceeding the normal use range and monitoring the time to failure. Data must be collected over a range of time from 1 h to beyond 10,000 hours. The pressure/stress/strain (all may be validly used) versus time to failure is statistically extrapolated to 50 years to establish a long-term HDB. To establish the pipe pressure rating, a safety factor is applied to this 50-year value. In ANSI/AWWA C950, Fiberglass Pressure Pipe, the specified safety factor is 1.8 at 50 years (i.e., the extrapolated 50-year value is divided by 1.8 to establish the allowable design value).

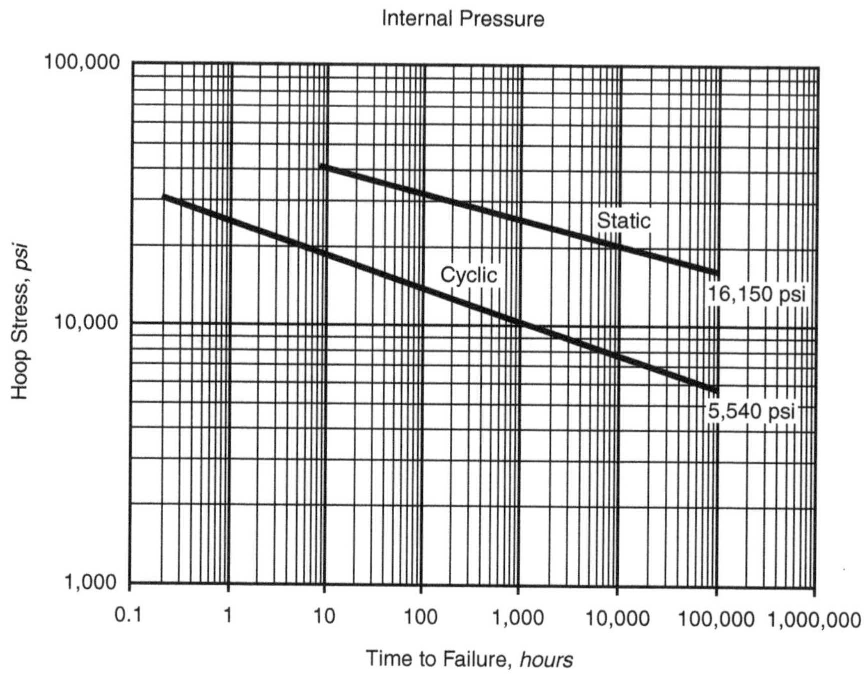

Figure 2-3 Static versus cyclic pressure testing

This testing may be conducted using static pressurization (the standard for water piping) or cyclic pressure testing (which is common for small-diameter pipes used in the oil-field industry). The same pipe tested in both static and cyclic pressure conditions will exhibit significantly different regression behavior. The cycling testing condition is far more severe (25 cycles per min from 0 to test pressure). Because the test is so severe, the common practice is to use the 50-year value directly for design purposes (i.e., the safety factor applied to the extrapolated 50-year value is 1.0). To illustrate the comparison of the two procedures, Figure 2-3 shows the results of a filament-wound epoxy pipe tested by both static and cyclic pressure testing procedures.

2.7 REFERENCES

ANSI/AWWA (American National Standards Institute/American Water Works Association). 2020. C950. Fiberglass Pressure Pipe. Denver: AWWA.

ASTM (ASTM International). 2022. D2992. Standard Practice for Obtaining Hydrostatic or Pressure Design Basis for "Fiberglass" (Glass-Fiber-Reinforced Thermosetting-Resin) Pipe and Fittings. West Conshohocken, PA: ASTM.

Chapter 3

Manufacturing

3.1 INTRODUCTION

Machine-made fiberglass pipe is produced using two basic processes: filament winding and centrifugal casting. Each process produces a pipe with characteristics that, although unique and advantageous for some applications, will meet the performance requirements of ANSI/AWWA Standard C950, Fiberglass Pressure Pipe.

3.2 FILAMENT WINDING

Filament winding is the process of impregnating glass fiber reinforcement with resin then applying the wetted fibers onto a mandrel in a prescribed pattern. Fillers, if used, are added during the winding process. Chopped glass rovings may be used as supplemental reinforcement. Repeated application of wetted fibers, with or without filler, results in a multilayered structural wall construction of the required thickness. After curing, the pipe may undergo one or more auxiliary operations such as joint preparation. The inside diameter (ID) of the finished pipe is fixed by the mandrel outside diameter (OD). The OD of the finished pipe is variable and determined by the pipe wall thickness.

The filament winding process is illustrated in Figure 3-1. Within the broad definition of filament winding, several methods are used—including reciprocal, continuous, multiple-mandrel, and ring and oscillating mandrel—each of which is described briefly. Figure 3-2 shows the application of impregnated glass reinforcement onto a mandrel during production of a filament-wound pipe.

3.2.1 Reciprocal Method

The reciprocal method is the most widely used filament winding production method. In this method, the fiber placement head with the associated resin bath drives back and forth past a rotating mandrel (see Figure 3-1). The angle of fiber placement relative to the

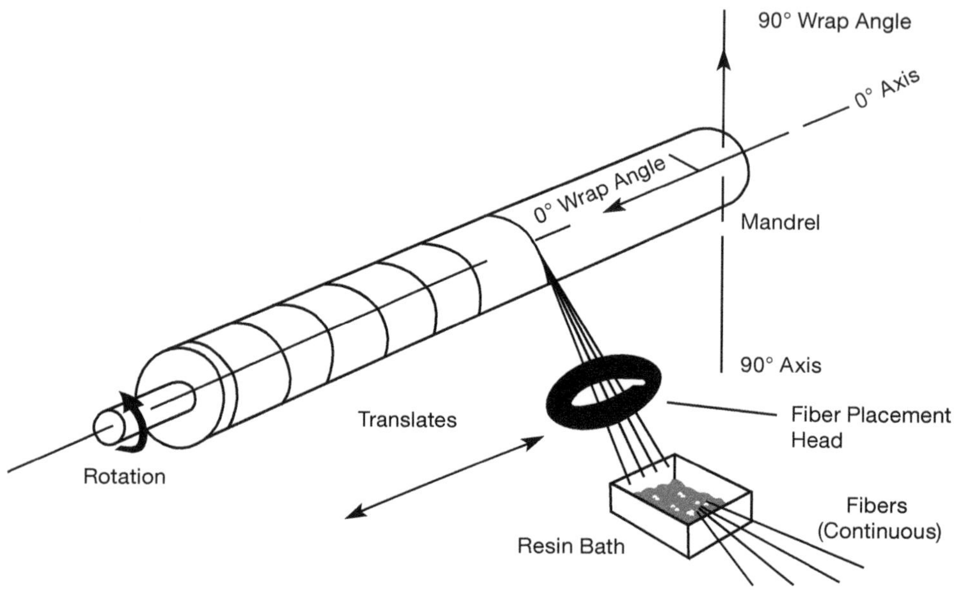

Source: Fiberglass Pipe Handbook, Fiberglass Pipe Institute, New York

Figure 3-1 Filament winding process

Figure 3-2 Application of impregnated glass reinforcement of a filament-wound pipe

mandrel axis is controlled by the synchronized translational speed of the bath and the rotational speed of the mandrel.

3.2.2 Continuous Methods

In one type of continuous process, pipe is made on one or more mandrels, which move past stations that apply fiberglass tapes preimpregnated with resin or glass fiber and resin. The winding angles are controlled through a combination of longitudinal mandrel speed, mandrel rotation (if used), or the rotation of planetary glass application stations. Once started, these methods produce pipe continuously, stopping only to replenish or change material components.

A second type of continuous process is the continuous advancing mandrel, which is composed of a continuous steel band supported by beams, which form a cylindrically shaped mandrel. The beams rotate, friction pulls the band around, and roller bearings allow the band to move longitudinally so that the entire mandrel continuously moves in a spiral path toward the end of the machine. Raw materials (continuous fibers, chopped fibers, resin, and aggregate fillers) are fed to the mandrel from overhead. Release films and surfacing materials are applied from rolls adjacent to the mandrel. After curing, a synchronized saw unit cuts the pipe to proper length. This method is illustrated in Figure 3-3. Finished pipe emerging from the curing oven is shown in Figure 3-4.

3.2.3 Multiple-Mandrel Method

In the multiple-mandrel method, a single materials-application system applies wetted glass reinforcement simultaneously to two or more mandrels. When the winding operation finishes, the mandrels are indexed to a new position for curing while another set of mandrels is wound.

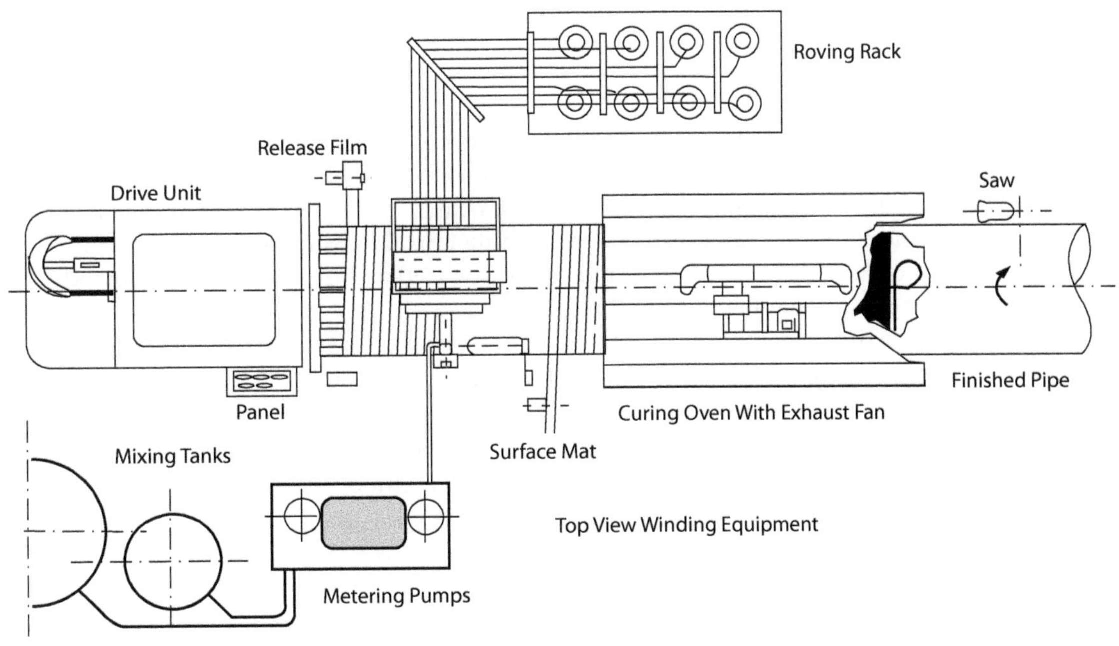

Source: Flowtite Technology, Sandefjord, Norway

Figure 3-3 Continuous advancing mandrel method

Source: Flowtite Technology, Sandefjord, Norway

Figure 3-4 Finished pipe emerging from curing oven

3.2.4 Ring and Oscillating Mandrel Method

The use of 360° glass delivery systems, sometimes in combination with an oscillating mandrel, allows production with both high- and low-winding angles as single circuit patterns (without interlayer crossovers).

3.3 CENTRIFUGAL CASTING

Centrifugal casting is used to manufacture tubular goods by applying resin and reinforcement to the inside of a mold that is rotated and heated, subsequently polymerizing (curing) the resin system. The OD of the finished pipe is determined by the ID of the mold tube. The ID of the finished pipe is variable and determined by the amount of material introduced into the mold. Other materials, such as sand or fillers, may be introduced in the process during manufacture of the pipe.

Two different methods of centrifugal casting are used and are described briefly.

1. *Preformed glass reinforcement sleeve method.* A preformed glass reinforcement sleeve is placed inside a steel mold. As the steel mold rotates, resin and a filler, if used, are placed within the mold by a feed tube that moves in and out of the mold, thus wetting out the preformed sleeve. This method is illustrated in Figure 3-5.
2. *Chopped glass reinforcement method.* Varying proportions of chopped glass reinforcement, resin, and aggregate are introduced simultaneously, by layer, from a feeder arm that moves in and out of the mold. This method is illustrated in Figure 3-6. Application of glass, resin, and sand within a rotating mold is shown in Figure 3-7.

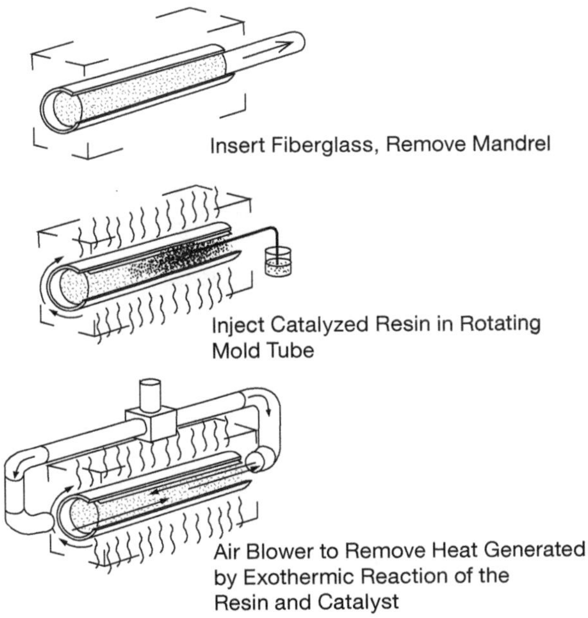

Figure 3-5 Preformed glass reinforcement sleeve method

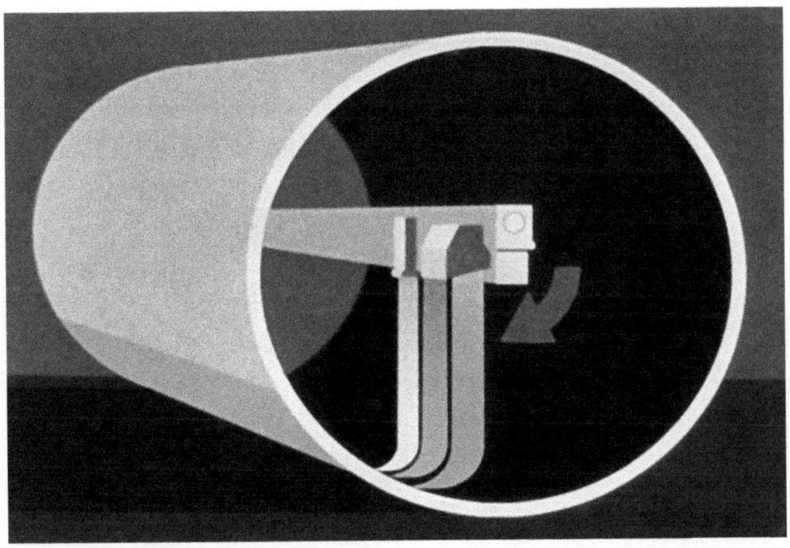

Source: Hobas Pipe USA Inc., Houston, Tex.

Figure 3-6 Chopped glass reinforcement method

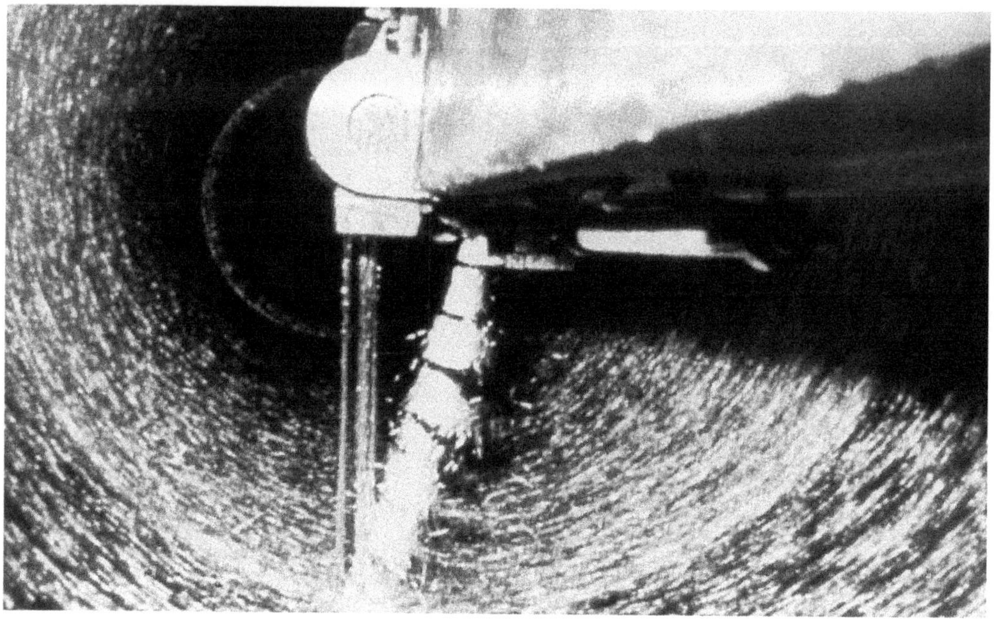

Source: Hobas Pipe USA Inc., Houston, Tex.

Figure 3-7 Application of glass, resin, and sand

3.4 REFERENCE

ANSI/AWWA (American National Standards Institute/American Water Works Association). 2020. C950. Fiberglass Pressure Pipe. Denver: AWWA.

Chapter 4

Hydraulics

4.1 HYDRAULIC CHARACTERISTICS

The smooth interior of fiberglass pipe results in low fluid resistance, which could lower horsepower requirements for pumped systems. Because the interior pipe surface typically remains smooth over time in most fluid services, fluid resistance does not increase with age. In addition, the smooth interior allows the pipe diameter to be reduced while maintaining the desired flow.

This chapter provides a basis for analysis of the flow capacity, economics, and fluid transient characteristics of fiberglass pipe.

4.2 PRELIMINARY PIPE SIZING

The first step in designing a piping system is to determine the pipe size needed to transport a specific amount of fluid. Many engineers have adopted rules that are independent of pipe length but rely on typical or limiting fluid velocities or allowable pressure loss per 100 ft (30 m) of pipe. Once the fluid velocity or the pressure loss is known, a pump can be sized to provide the proper flow rate at the required pressure. The following equations are guidelines for the initial sizing of pipe. These equations are presented with inch-pound units in the left-hand column and metric units in the right-hand column.

4.2.1 Maximum Velocity

$V = 48/\rho^{0.33}$

Where:

V = fluid velocity, ft/s
ρ = fluid density, lb/ft^3
= 62.4 lb/ft^3 for water

$V = 37/\rho^{0.33}$ (Eq 4-1)

Where:

V = fluid velocity, m/s
ρ = fluid density, kg/m^3
= 1,000 kg/m^3 for water

4.2.2 Maximum Velocity for Corrosive or Erosive Fluids

$$V = 24/\rho^{0.33} \qquad V = 18.4/\rho^{0.33} \qquad \text{(Eq 4-2)}$$

4.2.3 Minimum Pipe Diameter for Water

$$d = 0.73[(Q)/(SG)]^{0.5}/\rho^{0.33} \qquad d = 186[(Q)(SG)]^{0.5}/\rho^{0.33} \qquad \text{(Eq 4-3)}$$

Where:

d = pipe inside diameter, in.
Q = flow rate, gallons per minute (gpm)
SG = specific gravity, dimensionless
= 1.0 for water

Where:

d = pipe inside diameter, mm
Q = flow rate, L/s
SG = specific gravity, dimensionless
= 1.0 for water

4.2.4 Minimum Pipe Diameter for Corrosive or Erosive Fluids

$$d = 1.03[(Q)/(SG)]^{0.5}/\rho^{0.33} \qquad d = 262[(Q)/(SG)]^{0.5}/\rho^{0.33} \qquad \text{(Eq 4-4)}$$

4.3 TYPICAL PIPE DIAMETERS

The equations in Sec. 4.2 represent the minimum pipe diameters or maximum fluid velocities for water and corrosive (or erosive) liquid flow. Typical diameters for fiberglass pressure pipe and suction pipe can be calculated using the following equations.

4.3.1 Typical Diameters for Pressure Pipe Service

$$d = 0.321[(Q)/(SG)^2]^{0.434} \qquad d = 27.1[(Q)/(SG)^2]^{0.434} \qquad \text{(Eq 4-5)}$$

4.3.2 Typical Diameters for Suction Pipe Service

$$d = 0.434[(Q)/(SG)^2]^{0.434} \qquad d = 36.6[(Q)/(SG)^2]^{0.434} \qquad \text{(Eq 4-6)}$$

4.3.3 Conversion of Flow Rate to Fluid Velocity

$$V = 0.409(Q/d^2) \qquad V = 1{,}274(Q/d^2) \qquad \text{(Eq 4-7)}$$

4.4 PRESSURE REDUCTION CALCULATIONS

Reduction in pressure, expressed as head loss in feet (meters) or pressure loss in pounds per square inch (kilopascals), occurs in all piping systems because of elevation changes, turbulence caused by abrupt changes of direction, and friction within the pipe and fittings.

Different computational methods can be used to determine the head loss in fiberglass pipe. The most common methods are the Hazen–Williams, Manning, and Darcy–Weisbach equations. The suitability of each method depends on the type of flow (gravity or pumped) and the level of accuracy required.

4.4.1 Hazen–Williams Equation

The Hazen–Williams equation is applicable to water pipes under conditions of full turbulent flow. Although not as technically correct as other methods for all velocities, the Hazen–Williams equation has gained wide acceptance in the water and wastewater industries.

The Hazen–Williams equation is presented in nomograph form in Figure 4-1, which is typical for small-diameter fiberglass pipe. Graphical solutions, however, are usually valid for water only. When fluids other than water are encountered, a more universal solution, such as the Darcy–Weisbach equation, should be used. The Hazen–Williams equation is valid for turbulent flow and will usually provide a conservative solution for determining the head loss in fiberglass pipe.

$$h_f = 0.2083\,(100/C)^{1.85}(Q^{1.85}/d^{4.87}) \qquad h_f = 240\times10^6\,(100/C)^{1.85}(Q^{1.85}/d^{4.87}) \qquad \text{(Eq 4-8)}$$

Where:

h_f = friction factor, ft of water/100 ft

C = Hazen–Williams roughness coefficient, dimensionless (typical value for fiberglass pipe = 150)

Where:

h_f = friction factor, m of water/100 m

C = Hazen–Williams roughness coefficient, dimensionless (typical value for fiberglass pipe = 150)

Note: Graphs and examples use nominal pipe size for simplicity. The actual inside diameter (ID) should be used in hydraulic calculations.

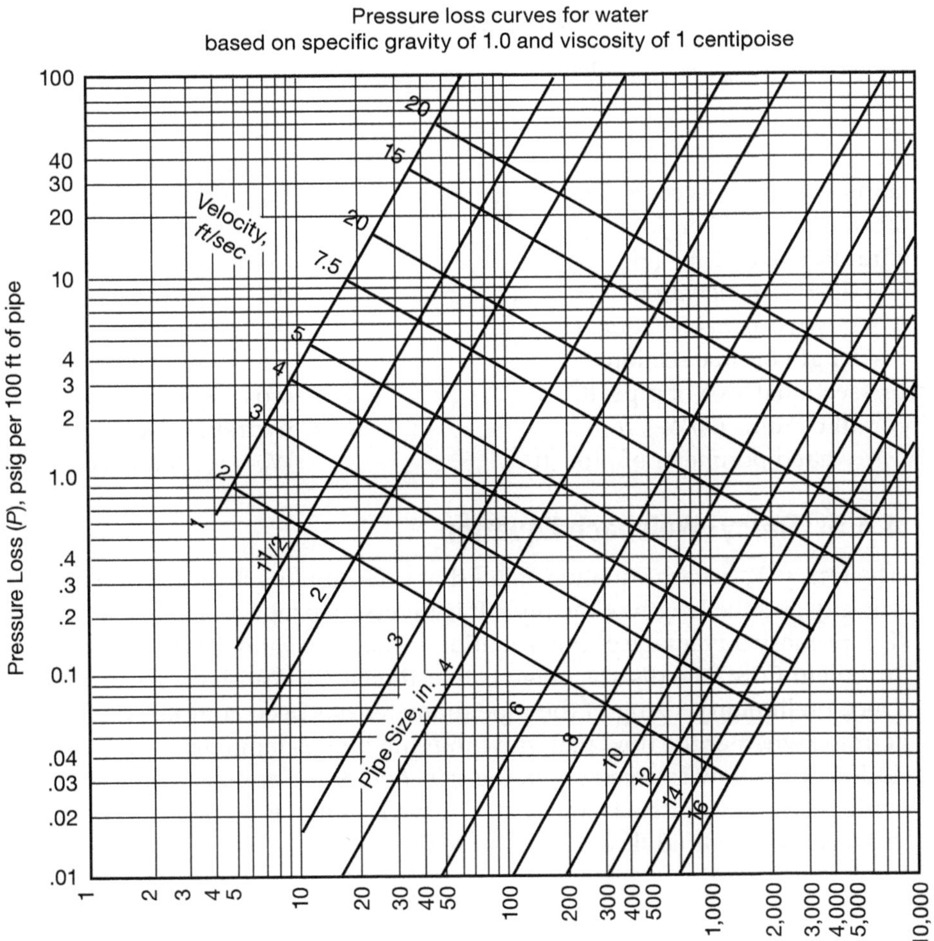

Figure 4-1 Friction pressure loss due to water flow through fiberglass pipe

4.4.2 Simplified Hazen–Williams

Many engineers prefer a simplified version of the Hazen–Williams equation:

$$h_f = \left[3.35 \times 10^6 Q/\left(Cd^{2.63}\right)\right]^{1.852} \qquad h_f = \left[3.35 \times 10^6 Q/\left(Cd^{2.63}\right)\right]^{1.852} \qquad \text{(Eq 4-9)}$$

4.4.3 Head Loss Converted to Pressure Loss

Head loss for any liquid is converted into pressure loss using the following equation:

$$p = \left(H_f\right)(SG)/2.31 \qquad p = \left(H_f\right)(SG)/0.102 \qquad \text{(Eq 4-10)}$$

Where:

p = pressure loss, psi
H_f = $(h_f)(L)/100$, ft
L = line length, ft

Where:

p = pressure loss, kPa
H_f = $(h_f)(L)/100$, m
L = line length, m

New fiberglass pipe has a Hazen–Williams roughness coefficient C value of 150–165. A design value of 150 is frequently used with fiberglass pipe.

4.4.4 Manning Equation

The Manning equation typically solves gravity flow problems in which the pipe is only partially full and is under the influence of an elevation head only.

$$Qm = (1.486/n)(S)^{0.5}(A)(R)^{0.667} \qquad Qm = (1{,}000/n)(S)^{0.5}(A)(R)^{0.667} \qquad \text{(Eq 4-11)}$$

Where:

Qm = flow rate, ft^3/s
n = Manning roughness coefficient
= 0.009 for typical fiberglass pipe
S = hydraulic slope, ft/ft = $(H_1 - H_2)/L$
H_1 = upstream elevation, ft
H_2 = downstream elevation, ft
L = length of pipe section, ft
A = cross-sectional area of pipe, ft^2
R = hydraulic radius (A/W_p), ft
W_p = wetted perimeter of pipe, ft

Where:

Qm = flow rate, L/s
n = Manning roughness coefficient
= 0.009 for typical fiberglass pipe
S = hydraulic slope, m/m = $(H_1 - H_2)/L$
H_1 = upstream elevation, m
H_2 = downstream elevation, m
L = length of pipe section, m
A = cross-sectional area of pipe, m^2
R = hydraulic radius (A/W_p), m
W_p = wetted perimeter of pipe, m

4.4.5 Darcy–Weisbach Equation

The Darcy–Weisbach equation states that pressure loss is proportional to the square of the velocity and the length of the pipe. It is inversely proportional to the diameter of the pipe. The primary advantage of this equation is that it is valid for all fluids in both laminar and turbulent flow. The disadvantage is that the Darcy–Weisbach friction factor is a variable. Once preliminary sizing of the pipe diameter has been completed, the next step is to determine whether the flow pattern within the pipe is laminar or turbulent. This characterization of the flow is necessary in the selection of the appropriate friction factor to be used with the Darcy–Weisbach equation. The well-known Reynolds number equation is used to characterize the fluid flow:

$$R_e = (ID)(V)/\mu \qquad R_e = (ID)(V)/\mu \qquad \text{(Eq 4-12)}$$

Where:

R_e = Reynolds number, dimensionless
ID = pipe inside diameter, ft
V = fluid velocity, ft/s
μ = kinematic viscosity, ft^2/s

Where:

R_e = Reynolds number, dimensionless
ID = pipe inside diameter, m
V = fluid velocity, m/s
μ = kinematic viscosity, m^2/s

This guideline determines the type of flow from the Reynolds number:

Flow Type	Reynolds Number
Laminar flow	$R_e \leq 2{,}000$
Transition flow zone	$2{,}000 < R_e < 4{,}000$
Turbulent flow	$R_e \geq 4{,}000$

Simply stated, the Darcy–Weisbach equation is as follows:

$$H_f = \left[fL\left(V^2\right)\right] / 2(ID)g \qquad H_f = \left[fL\left(V^2\right)\right] / 2(ID)g \qquad \text{(Eq 4-13)}$$

Where:
- f = Darcy–Weisbach friction factor, dimensionless
- g = gravitational constant = 32.2 ft/s^2

Where:
- f = Darcy–Weisbach friction factor, dimensionless
- g = gravitational constant = 9.81 m/s^2

If flow in the pipe is laminar (i.e., $R_e \leq 2{,}000$), the friction factor f_l reduces to

$$f_l = 64 / R_e \qquad f_l = 64 / R_e \qquad \text{(Eq 4-14)}$$

Where:
- f_l = friction factor for laminar flow

Where:
- f_l = friction factor for laminar flow

NOTE: Friction factor for laminar flow is denoted as f_l, and f_t denotes friction factor for turbulent flow.

When the flow regime is turbulent (i.e., $R_e \geq 4{,}000$), the friction factor can be determined from the Moody diagram, which is found in most fluid mechanics texts (Figure 4-2). Fiberglass pipe has a surface roughness parameter e equal to 1.7×10^{-5} ft (5.18×10^{-6} m). When divided by the pipe diameter (e/ID), the friction factor f_t for turbulent flow can be extracted from the smooth pipe segment of the diagram. The friction factor for turbulent flow can also be calculated from the Colebrook equation:

$$1/f_t^{0.5} = -2 \log\left[(e/\text{ID})/3.7\right] + 2.51/\left[(R_e)\left(f_t^{0.5}\right)\right] \qquad 1/f_t^{0.5} = -2 \log\left[(e/\text{ID})/3.7\right] + 2.51/\left[(R_e)\left(f_t^{0.5}\right)\right] \qquad \text{(Eq 4-15)}$$

Where:
- f_t = Moody friction factor, dimensionless
- e = surface roughness factor, ft
- = 1.7×10^{-5} as typical for fiberglass pipe
- ID = pipe inside diameter, ft

Where:
- f_t = Moody friction factor, dimensionless
- e = surface roughness factor, m
- = 5.18×10^{-6} as typical for fiberglass pipe
- ID = pipe inside diameter, m

This equation is difficult to solve because it is implicit in f_t and requires a trial-and-error iterative solution. The following simplified equation relates the friction factor to the Reynolds number and is accurate to within 1 percent of the Colebrook equation:

$$f_t = \left[1.8 \log\left(R_e / 7\right)\right]^{-2} \qquad f_t = \left[1.8 \log\left(R_e / 7\right)\right]^{-2} \qquad \text{(Eq 4-16)}$$

4.5 HEAD LOSS FITTINGS

Head loss in fittings is frequently expressed as the equivalent length of pipe that is added to the straight run of pipe. This approach has sufficient accuracy for many applications and is used most often with the Hazen–Williams or Manning equations. The approach does not consider turbulence and subsequent losses created by different fluid velocities. When tabular data are not available or when additional accuracy is necessary, head loss in fittings (or valves) can be determined using loss coefficients (K factors) for each type of fitting. Table 4-1 provides the typical K factors. In this approach, the K factor for each fitting is multiplied by the velocity head of the fluid flow. Eq 4-17 illustrates the loss coefficient approach.

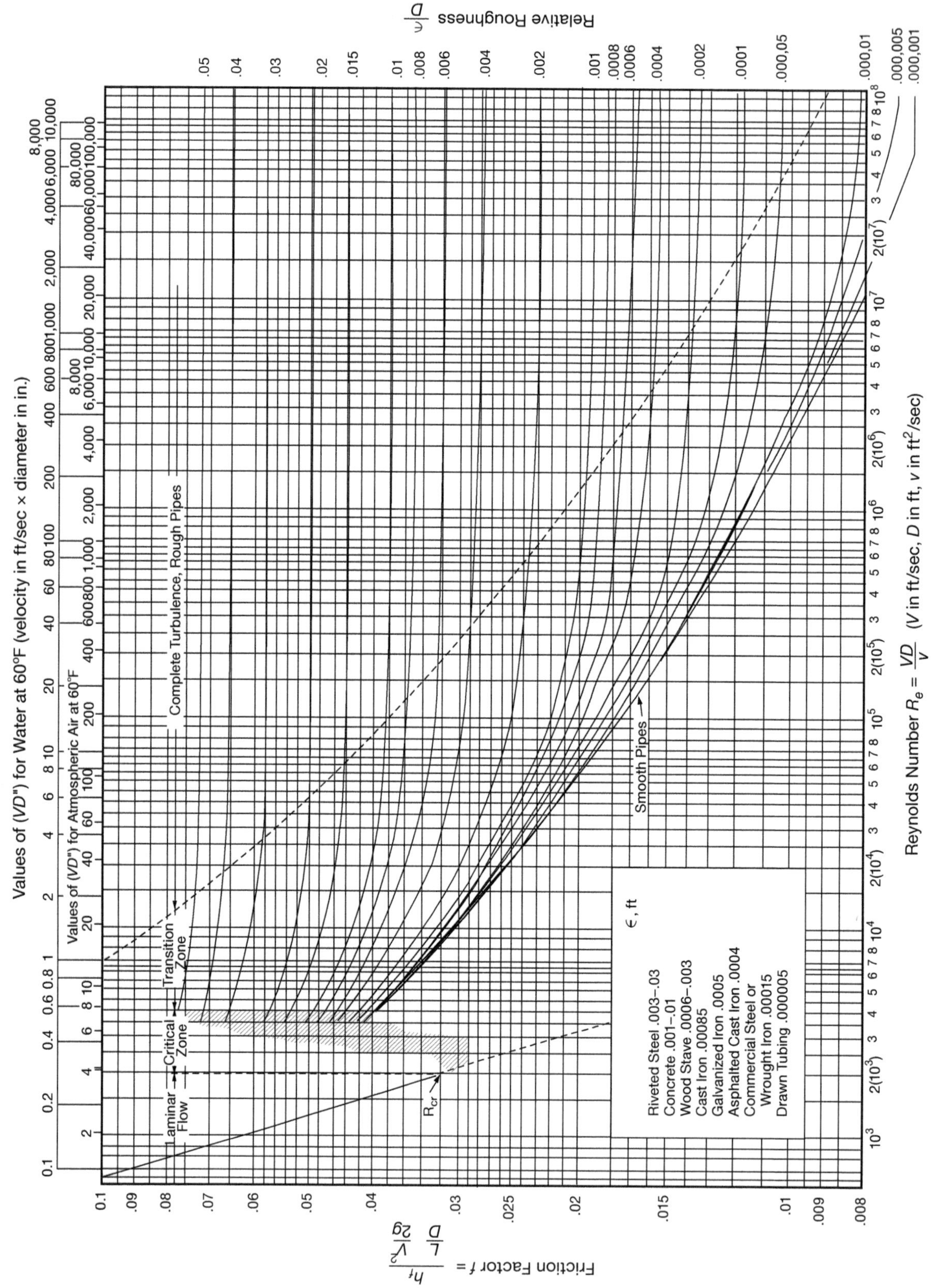

Source: L.F. Moody, *Friction Factors for Pipe Flow*, ASME, 345 E. 47th St., New York, NY 10017

Figure 4-2 Moody diagram for determination of friction factor for turbulent flow

Table 4-1 Typical *K* factors for fiberglass fittings

Type of Fitting	*K* Factor
90° elbow, standard	0.5
90° elbow, single miter	1.4
90° elbow, double miter	0.8
90° elbow, triple miter	0.6
180° return bend	1.3
Tee, straight flow	0.4
Tee, flow to branch	1.4
Tee, flow from branch	1.7
Reducer, single-size reduction	0.7
Reducer, double-size reduction	3.3

$$H_{ff} = K\left(V^2/2g\right) \qquad H_{ff} = K\left(V^2/2g\right) \qquad \text{(Eq 4-17)}$$

Where:
- H_{ff} = equivalent fittings head loss, ft
- K = *K* factor for each fitting type from Table 4-1

Where:
- H_{ff} = equivalent fittings head loss, m
- K = *K* factor for each fitting type from Table 4-1

Many hydraulic handbooks provide *K* factors for various types of fittings and valves not included in this manual.

The total head loss in a system includes but is not limited to losses from fittings, the head loss from the straight run pipe, and head losses due to changes in elevation.

4.6 ENERGY CONSUMPTION CALCULATION PROCEDURE

Pipeline operating costs are driven, in large part, by the frictional resistance of the pipe and the corresponding power consumption, and the pipeline design process should consider the operating economics of the pipeline material. This section outlines the basic procedure for determining the head loss because of friction and relative economic merits when considering different pipe materials.

4.6.1 Pipeline Economic Analysis Procedure

Step 1. Calculate the head loss (Eq 4-9):

$h_f = [42.7\ Q/(Cd^{2.63})]1.852$ $\qquad$ $h_f = [3.35 \times 10^6\ Q/(Cd^{2.63})]1.852$

Step 2. Convert head loss to pump horsepower demand:

$$HP = Q\rho_1 H_f / 33{,}000 \qquad HP = Q\rho_1 H_f / 102 \qquad \text{(Eq 4-18)}$$

Where:
- HP = water-pumping power required, hp; 1 hp = 33,000 ft-lbf/min
- ρ_1 = fluid density = lb/gal
  = 8.34 lb/gal for water
- H_f = $(h_f)(L)/100$, ft
- L = line length, ft

Where:
- HP = water-pumping power required, kW; 1 kW = 102 kg-m/s
- ρ_1 = fluid density = kg/L
  = kg/L for water
- H_f = $(h_f)(L)/100$, m
- L = line length, m

Step 3. Calculate the annual energy usage:

(To demonstrate the calculations in a clear format, the following equations assume the pumps run 24 h/day at full capacity. This is not a realistic assumption. In design situations, engineers must assess the actual expected operating conditions, e.g., 10 percent of the time at full power, 25 percent of the time at 75 percent power.)

$$Ec = (HP)(24)(365)(0.7457)/\,eff$$

Where:

- Ec = annual energy consumption, kW•h per year
- 24 = hours per day
- 365 = days per year
- 0.7457 = kW•h/hp-h
- eff = overall (pump and motor) efficiency, typically, 75–85 percent

$$Ec = (HP)(24)(365)/\,eff \qquad \text{(Eq 4-19)}$$

Where:

- Ec = annual energy consumption, kW•h per year
- 24 = hours per day
- 365 = days per year
- eff = overall (pump and motor) efficiency, typically 75–85 percent

Step 4. Calculate average annual energy cost (AEC):

$$\text{AEC} = (Ec)(\text{UEC})$$

Where:

- AEC = annual energy cost, $
- UEC = unit energy cost, $/kW•h

$$\text{AEC} = (Ec)(\text{UEC}) \qquad \text{(Eq 4-20)}$$

Where:

- AEC = annual energy cost, $
- UEC = unit energy cost, $/kW•h

Techniques that consider the time value of money (e.g., net present value, life cycle costing) can also be used to evaluate the relative economics of alternative pipe materials. These techniques consider the installed cost of pipe in the calculation, and future cash flows are discounted to current value.

4.7 PRESSURE SURGE

Pressure surge, also known commonly as water hammer, results from an abrupt change of fluid velocity within the system. The pressure surge results from the rapidly moving wave that increases and decreases the pressure in the system depending on the source and direction of wave travel. Under certain conditions, pressure surges can reach magnitudes sufficient to rupture or collapse a piping system, regardless of the material of construction.

Rapid valve closure can result in the buildup of pressure waves due to the conversion of kinetic energy of the moving fluid to potential energy that must be accommodated. These pressure waves will travel throughout the piping system and can cause damage far away from the wave source.

4.7.1 Cause and Control of Pressure Surge

The magnitude of pressure surge is a function of the fluid properties and velocity, the modulus of elasticity and wall thickness of the pipe material, the length of the line, and the speed at which the momentum of the fluid changes. The relatively high compliance (low modulus of elasticity) of fiberglass pipe contributes to a self-damping effect as the pressure wave travels through the piping system.

In addition to rapid valve closure or opening, sudden air release and pump startup or shutdown can create pressure surge. Pressure surges do not show up readily on conventional Bourdon tube gauges because of the slow response of the instrument. The net result of pressure surge can be excessive pressures, pipe vibration, or movement that can cause failure in pipe and fittings.

In some cases, anchoring the piping system may mitigate pipe vibration and movement problems. In other cases, mechanical valve operators, accumulators, rupture discs, surge relief valves, feedback loops around pumps, etc., may be required to protect against or remove the source of pressure surge.

Good design practice usually prevents pressure surge in most systems. Installation of valves that cannot open or close rapidly is one simple precaution. In addition, pumps should never be started in empty discharge lines unless slow-opening, mechanically actuated valves can increase the flow rate gradually.

4.7.2 Pressure Surge Calculation

The Talbot equation calculates surge pressure due to a change in velocity:

$$P_s = (a/g)(SG/2.3)(\Delta V) \qquad P_s = (a)(SG)(\Delta V) \qquad \text{(Eq 4-21)}$$

Where:

$$a = \frac{12}{\left[\left(\frac{\rho}{g}\right)\left(\frac{1}{k} + \frac{d}{Et}\right)\right]^{0.5}}$$

P_s = pressure surge deviation from normal, psig
SG = fluid specific gravity, dimensionless
ΔV = change in flow velocity, ft/s
a = wave velocity, ft/s
ρ = fluid density, lb/ft^3
g = gravitational constant, 32.2 ft/s^2
k = bulk modulus of compressibility of liquid, psi
= 300,000 psi for water
d = pipe inside diameter, in.
E = modulus of elasticity of pipe wall, psi
t = pipe wall thickness, in.

Where:

$$a = \frac{1}{\left[\rho\left(\frac{1}{10^9 k} + \frac{d}{10^9 Et}\right)\right]^{0.5}} \qquad \text{(Eq 4-22)}$$

P_s = pressure surge deviation from normal, kPa
SG = fluid specific gravity, dimensionless
ΔV = change in flow velocity, m/s
a = wave velocity, m/s
ρ = fluid density, kg/m^3
k = bulk modulus of compressibility of liquid, GPa
= 2 GPa for water
d = pipe inside diameter, mm
E = modulus of elasticity of pipe wall, GPa
t = pipe wall thickness, mm

The pressure class P_c must be greater than or equal to the sum of the working pressure P_w and surge pressure P_s divided by 1.4 (see Chapter 5, Sec. 5.7.1.3).

Many fluid mechanics and hydraulic handbooks provide procedures such as the previous Talbot equation for calculating pressure surges as a result of a single valve closure in simple piping systems. Sophisticated fluid transient computer programs are also available to analyze pressure surge in complex multibranch piping systems under a variety of conditions.

4.8 DESIGN EXAMPLES

Example 4-1: Use of the Hazen–Williams equation

Compute the frictional pressure loss in a 1,500 ft–long, 10 in.–diameter fiberglass pipe transporting 2,000 gpm of water.

(C = 150 as typical)

Compute the frictional pressure loss in a 450 m–long, 250 mm–diameter fiberglass pipe transporting 130 L/s of water.

(C = 150 as typical)

Step 1. *Compute the head loss per unit length of pipe using Eq 4-9:*

h_f = $[42.7\ Q/(Cd^{2.63})]^{1.852}$
h_f = $[42.7(2{,}000)/(150)(10)^{2.63}]^{1.852}$
= 1.71 ft water/100 ft

h_f = $[3.35 \times 10^6\ Q/(Cd^{2.63})]^{1.852}$
h_f = $[3.35 \times 10^6\ (130)/(150)(250^{2.63})]^{1.852}$
= 1.95 m water/100 m

The total head loss for the length of pipe in this example is then:

H_f = 1.71 (1,500)/100 = 25.65 ft of water

H_f = 1.95 (450)/100 = 8.78 m of water

Step 2. *Convert head loss to pressure drop using Eq 4-10:*

p = $(H_f)(SG)/2.31$
= 25.65(1.0)/2.31 = 11.1 psi

p = $(H_f)(SG)/0.102$
= 8.78(1.0)/0.102 = 86 kPa

Example 4-2: Determine the pipe diameter, working pressure, and pressure class on a pipeline

A pipeline requires 5,000 ft of pipe, four 90-degree elbows (double miter), and three 45-degree elbows (single miter) with an elevation change of 25 ft. Assume the kinematic viscosity μ = .00001. The flow rate is 8,000 gpm.

A pipeline requires 1,500 m of pipe, four 90-degree elbows (double miter), and three 45-degree elbows (single miter) with an elevation change of 7.5 m. Assume kinematic viscosity $\mu = 0.93 \times 10^{-6}$. The flow rate is 500 L/s.

Step 1. *Determine minimum diameter (Eq 4-3):*

d = $0.73\ [(Q)/(SG)]^{0.5}/\rho^{0.33}$
d = $0.73\ [(8{,}000)/(1)]^{0.5}/62.4^{0.33}$
= 16.7 in.

d = $186\ [(Q)/(SG)]^{0.5}/\rho^{0.33}$
d = $186\ [(500)/(1)]^{0.5}/(1{,}000)^{0.33}$
= 425 mm

Therefore, use the next larger available size, which is 18 in. for purposes of this example (ID 18 in. = 1.5 ft).

Therefore, use the next larger available size, which is 450 mm for purposes of this example (ID 450 mm = .45 m).

Step 2. *Calculate average fluid velocity (Eq 4-7):*

V = $0.409\ (Q/d^2)$
= $0.409\ (8{,}000/(18)^2)$
= 10.1 ft/s

V = $1{,}274\ (Q/d^2)$
= $1{,}274\ (500/(450)^2)$
= 3.15 m/s

Step 3. *Calculate the Reynolds number (Eq 4-12):*

R_e = $(ID)\ (V)/\mu$
= (1.5) (10.1)/0.00001
= 1.52×10^6

R_e = $(ID)\ (V)/\mu$
= $(0.45)\ (3.15)/0.93 \times 10^{-6}$
= 1.48×10^6

Because R_e is greater than 4,000, flow is turbulent.

Step 4. *Calculate the friction factor (Eq 4-16):*

f_t = $[1.8 \log (R_e/7)]^{-2}$
= $[1.8 \log (1.52 \times 10^6/7)]^{-2}$
= 0.0108

f_t = $[1.8 \log (R_e/7)]^{-2}$
= $[1.8 \log (1.48 \times 10^6/7)]^{-2}$
= 0.0109

Step 5. Calculate system friction loss using Eq 4-13 and Eq 4-17:

Use a K factor for 90-degree double-miter elbows of 0.8 and for 45-degree single-miter elbows of 0.5. Consequently, the total K factor is 4(0.8) + 3 (0.5) = 4.7.

$$H_f = K(V^2/2g) + [f_t\, L(V^2)]/(\text{ID})2g$$
$$= V^2/2g\,[K + f_t\, L/\text{ID}]$$
$$= (10.1)^2/2(32.2)\,[4.7 + [0.0108\,(5{,}000)]/(1.5)]$$
$$= 64.5 \text{ ft}$$

$$H_f = K(V^2/2g) + [f_t\, L(V^2)]/(\text{ID})2g$$
$$= V^2/2g\,[K + f_t\, L/\text{ID}] = (3.15)^2/2(9.81)\,[4.7 + [0.0109\,(1{,}500)]/(0.45)]$$
$$= 20.8 \text{ m}$$

Step 6. Combine friction and elevation head:

$$H_{total} = 64.5 + 25 = 89.5 \text{ ft}$$

$$H_{total} = 20.8 + 7.5 = 28.3 \text{ m}$$

Step 7. Convert head loss to working pressure (Eq 4-10):

$$p = (H_f)(\text{SG})/2.31$$
$$= (89.5)\,(1.0)/2.31 = 38.7 \text{ psi}$$

$$p = (H_f)(\text{SG})/0.102$$
$$= (28.3)\,(1.0)/0.102 = 277 \text{ kPa}$$

The total working pressure of 38.7 psi would require a 50-psi pressure class. However, a higher pressure class may tentatively be selected to account for possible water hammer in the line. For these examples, assume that a 100-psi class is selected. See example 4-4 to verify that this is adequate for pressure surge.

The total working pressure of 277 kPa would require a 350-kPa pressure class. However, a higher pressure class may tentatively be selected to account for possible water hammer in the line. For these examples, assume that a 700-kPa class is selected. See example 4-4 to verify that this is adequate for pressure surge.

Example 4-3: Comparative power cost calculation

Assume a 10,000 ft–long, 6 in.–diameter pipeline is to deliver 500 gpm of water on a year-round basis. The engineer is considering using fiberglass pipe with an average Hazen–Williams coefficient $C = 150$ and another material that will have an average Hazen–Williams coefficient $C = 100$ over the life of the pipeline. Calculate the average AEC for each candidate material and the total energy cost over a 20-year service life using a unit cost of power of \$0.06/kW•h.

Assume a 3,000 m–long, 150 mm–diameter pipeline is to deliver 30 L/s of water on a year-round basis. The engineer is considering using fiberglass pipe with an average Hazen–Williams coefficient $C = 150$ and another material that will have an average Hazen–Williams coefficient $C = 100$ over the life of the pipeline. Calculate the average AEC for each candidate material and the total energy cost over a 20-year service life using a unit cost of power of \$0.06/kW•h.

Step 1. Calculate the head loss for each material (Eq 4-9):

$$h_f = [42.7\, Q/(Cd^{2.63})]^{1.852}$$

$$h_f = [3.35 \times 10^6\, Q/(Cd^{2.63})]^{1.852}$$

For fiberglass pipe:

$$h_f = [42.7 \times 500/(150 \times 6^{2.63})]^{1.852}$$
$$= 1.58 \text{ ft/100 ft}$$
$$H_f = 158 \text{ ft for the entire line}$$

$$h_f = [3.35 \times 10^6 \times 30/(150 \times 150^{2.63})]^{1.852}$$
$$= 1.55 \text{ m/100 m}$$
$$H_f = 46.5 \text{ m for the entire line}$$

For alternate material:

$= [42.7 \times 500/(100 \times 6^{2.63})]^{1.852}$ | $= [3.35 \times 10^6 \times 30/(100 \times 150^{2.63})]^{1.852}$

= 3.34 ft/100 ft | = 3.29 m/100 m

= 334 ft for the entire line | = 98.6 m for the entire line

Step 2. *Convert head loss to horsepower demand (Eq 4-18):*

$HP = Q \rho_1 H_f/33{,}000$ | $HP = Q \rho_1 H_f/102$

For fiberglass pipe:

= 500(8.34)(158)/33,000 | = 30(1)(46.5)/102

= 19.96 hp | = 13.68 kW

For alternate material:

= 500(8.34)(334)/33,000 | = 30(1)(98.6)/102

= 42.20 hp | = 29 kW

Step 3. *Calculate the annual energy consumption* Ec *using an 80 percent overall pump efficiency and Eq 4-19:*

(As previously noted, for simplicity in demonstrating the calculation procedure, this example assumes constant pump operation at full power. In design situations, engineers must assess actual operating levels.)

$Ec = (HP)(24)(365)(0.7457)/eff$ | $Ec = (HP)(24)(365)/eff$

For fiberglass pipe:

$= (HP)(24)(365)(0.7457)/eff$ | $= (HP)(24)(365)/eff$

= 19.96(24)(365)(0.7457)/0.80 | = 13.68(24)(365)/0.80

= 163,000 kW•h | = 150,000 kW•h

For alternate material:

$= (HP)(24)(365)(0.7457)/eff$ | $= (HP)(24)(365)/eff$

= 42.20(24)(365)(0.7457)/0.80 | = 29(24)(365)/0.80

= 344,620 kW•h | = 318,000 kW•h

Step 4. *Calculate the AEC (Eq 4-20) and calculate the total energy cost over 20 years:*

AEC = (*Ec*) (UEC) | AEC = (*Ec*) (UEC)

For fiberglass pipe:

= 163,000 (0.06) = \$9,780/year | = 150,000 (0.06) = \$9,000/year

= \$195,600 over 20 years | = \$180,000 over 20 years

For alternate material:

= 344,620 (0.06) = \$20,676/year | = 318,000 (0.06) = \$19,080/year

= \$413,000 over 20 years | = \$381,600 over 20 years

Example 4-4: Surge-pressure calculation

Determine if the maximum pressure surge for the pipe in example 4-2 is within the 40 percent surge allowance criteria. Assume a full instantaneous change in velocity equal to the flow velocity in the pipe. The fiberglass pipe has a tensile modulus of 3×10^6 psi and a pressure class of 100 psi. The pipe wall thickness $t = 0.21$ in. The bulk modulus of water is 300,000 psi.

Determine if the maximum pressure surge for the pipe in example 4-2 is within the 40 percent surge allowance criteria. Assume a full instantaneous change in velocity equal to the flow velocity in the pipe. The fiberglass pipe has a tensile modulus of 20 GPa and a pressure class of 700 kPa. The pipe wall thickness $t = 5.3$ mm. The bulk modulus of water is 2.07 GPa.

Step 1. *Calculate the wave velocity (Eq 4-22):*

$$a = 12/[(\rho/g)\ (1/k + d/(E(t)))]^{0.5}$$
$$= 12/[(62.4/32.2)\ (1/(3 \times 10^5) + 18/3 \times 10^6\ (0.21))]^{0.5}$$
$$= 1{,}526 \text{ ft/s}$$

$$a = 1/[(\rho)\ (1/10^9k + d/(10^9E(t)))]^{0.5}$$
$$= 1/[1{,}000(1/10^9(2) + 450/10^9(20)(5.3))]^{0.5}$$
$$= 459 \text{ m/s}$$

Step 2. *Calculate the surge pressure (Eq 4-21):*

$$P_s = (a/g)\ (SG/2.3)\ (\Delta V)$$
$$= (1{,}526/32.2)(1/2.3)(10.1)$$
$$= 208 \text{ psi}$$

$$P_s = (a)\ (SG)\ (\Delta V)$$
$$= (459)(1)(3.15)$$
$$= 1{,}446 \text{ kPa}$$

Step 3. *Check compliance with the maximum system pressure requirement:*

$$P_c \geq (P_w + P_s)/1.4$$

$$P_c \geq (P_w + P_s)/1.4$$

From example 4-2, P_w = 38.7 psi and the selected pressure class was P_c = 100 psi:

$$(38.7 + 208)/1.4 = 176 \text{ psi}$$

This exceeds the pressure class. The engineer has three options. The first would be to increase the pressure class to accommodate the surge, maintaining the same pipe diameter. The second would be to increase pipe diameter, which, together with a more moderate increase in pressure class, would satisfy the maximum system pressure requirement. Increasing the pipe diameter will decrease the fluid velocity, which reduces the friction loss, thereby lowering the working pressure. The third option is to provide measures, such as a surge tank, to reduce the magnitude of the surge.

For this example, the second option will be used, and a diameter of 20 in. (wall thickness of 0.23 in.) with a pressure class of 150 psi will be evaluated.

From example 4-2, P_w = 277 kPa and the selected pressure class was P_c = 700 kPa:

$$(277 + 1{,}446)/1.4 = 1{,}231 \text{ kPa}$$

This exceeds the pressure class. The engineer has three options. The first would be to increase the pressure class to accommodate the surge, maintaining the same pipe diameter. The second would be to increase pipe diameter, which together with a more moderate increase in pressure class would satisfy the maximum system pressure requirement. Increasing the pipe diameter will decrease the fluid velocity, which reduces the friction loss, thereby lowering the working pressure. The third option is to provide measures, such as a surge tank, to reduce the magnitude of the surge.

For this example, the second option will be used, and a diameter of 500 mm (wall thickness of 5.8 mm) with a pressure class of 1,000 kPa will be evaluated.

Step 4. *Calculate the fluid velocity for the new pipe diameter (Eq 4-7):*

$$V = 0.409(Q/d^2)$$
$$= 0.409[8{,}000/(20)^2]$$
$$= 8.18 \text{ ft/s}$$

$$V = 1{,}274(Q/d^2)$$
$$= 1{,}274[500/(500)^2]$$
$$= 2.55 \text{ m/s}$$

NOTE: This velocity is lower than the 10.1 ft/s (3.15 m/s) in example 4-2.

Step 5. *Calculate the new working pressure:*

A. Reynolds number (Eq 4-12):

$$R_e = (ID)\ (V)/\mu$$
$$= (20/12)\ (8.18)/0.00001$$
$$= 1{,}363{,}333$$

$$R_e = (ID)\ (V)/\mu$$
$$= (500/1{,}000)\ (2.55)/0.93 \times 10^{-6}$$
$$= 1{,}370{,}967$$

B. Friction factor (Eq 4-16):

$f_t = [1.8 \log (R_e/7)]^{-2}$
$= [1.8 \log (1{,}363{,}333/7)]^{-2}$
$= 0.01103$

$f_t = [1.8 \log (R_e/7)]^{-2}$
$= [1.8 \log (1{,}370{,}967/7)]^{-2}$
$= 0.01102$

C. Friction losses (using Eq 4-13 and Eq 4-17):

Use a K factor for 90-degree double-miter elbows of 0.8 and for 45-degree single-miter elbows of 0.5.

The total K factor is then 4(0.8) + 3(0.5) = 4.7.

$H_f = K (V^2/2g) + [f_t L(V^2)]/2(\text{ID})g$
$= V^2/2g\,[K + f_t L/\text{ID}]$
$= (8.18)^2/2(32.2)\,[4.7 +$
$[0.01103 (5{,}000)]/(20/12)]$
$= 39.3$ ft

$H_f = K (V^2/2g) + [f_t L(V^2)]/2(\text{ID})g$
$= V^2/2g\,[K + f_t L/\text{ID}]$
$= (2.55)^2/2(9.81)\,[4.7 +$
$[0.01102 (1{,}500)]/.5]$
$= 12.5$ m

Combine friction and elevation head:

$H_{total} = 39.3 + 25 = 64.3$ ft

$H_{total} = 12.5 + 7.5 = 20$ m

D. Convert to working pressure (Eq 4-10) and using H_{total} for H_f:

$p = (H_f)(\text{SG})/2.31$
$= (64.3)(1.0)/2.31 = 27.8$ psi

$p = (H_f)(\text{SG})/0.102$
$= (20)(1.0)/0.102 = 196$ kPa

Step 6. Calculate the wave velocity (Eq 4-22):

$a = 12/[(\rho/g)(1/k + d/(E(t)))]^{0.5}$
$= 12/[(62.4/32.2)(1/3 \times 10^5 +$
$20/3 \times 10^6 (0.23))]^{0.5}$
$= 1{,}516$ ft/s

$a = 1/[(\rho)(1/10^9 k + d/(10^9 E(t)))]^{0.5}$
$= 1/[(1{,}000(1/10^9(2) +$
$500/10^9(20)(5.8)))]^{0.5}$
$= 456$ m/s

Step 7. Calculate the pressure surge using (Eq 4-21):

$P_s = (a/g)(\text{SG}/2.3)(\Delta V)$
$= (1{,}516/32.2)(1/2.3)(8.18)$
$= 167$ psi

$P_s = (a)(\text{SG})(\Delta V)$
$= (456)(1)(2.55)$
$= 1{,}162$ kPa

Check compliance with maximum system pressure requirement:

$P_c \geq (P_w + P_s)/1.4$
150 psi $\geq (27.8 + 167)/1.4$
150 psi $\geq$ 39 psi

$P_c \geq (P_w + P_s)/1.4$
1,000 kPa $\geq (196 + 1{,}162)/1.4$
1,000 kPa $\geq$ 970 kPa

Therefore, the system pressure requirement is satisfied by using the higher pressure class in a larger diameter. Before final selection, the engineer would typically evaluate the economics of using the larger diameter with a higher pressure class versus using the original diameter with a still higher pressure class.

4.9 REFERENCES AND BIBLIOGRAPHY

ANSI/AWWA (American National Standards Institute/American Water Works Association). 2020. C950. Fiberglass Pressure Pipe. Denver: AWWA.

Benedict, R.P. 1980. *Fundamentals of Pipe Flow*. New York: John Wiley & Sons.

Brater, E.F., J. Lindell, C. Wei, and H.W. King. 1996 (7th ed.). *Handbook of Hydraulics*. New York: McGraw-Hill.

Fiberglass Pipe Institute. 1989. *Fiberglass Pipe Handbook.* New York: Fiberglass Pipe Institute.

Moody, L.F. 1944. "Friction Factors for Pipe Flow." *Transactions of the ASME,* 66:8:671–684.

Sharp, W.W., and T.M. Walski. 1988. "Predicting Internal Roughness in Water Mains." *Journal AWWA*. 80(11):34.

This page intentionally blank.

Chapter 5

Buried Pipe Design

5.1 INTRODUCTION

The structural design procedure for buried fiberglass pipe involves establishing design conditions, selecting pipe classes and corresponding pipe properties, selecting installation parameters, and performing pertinent calculations to ensure that the design requirements of Sec. 5.7 are satisfied. If the results of any calculation indicate that a requirement is not satisfied, it will be necessary to upgrade installation parameters or select a pipe with different properties, or both, and redo pertinent calculations. Special information and calculations not covered in this chapter may be required in unusual cases (see Sec. 5.9).

Both rigorous and empirical methods are used in the design of fiberglass pipe. In addition to short-term tests, many performance limits are determined at 50 years through statistical extrapolation of data obtained from long-term tests under simulated service conditions. Design stress or strain values are obtained by reducing performance limits using appropriate design factors. Design factors are established to ensure adequate performance over the intended service life of the pipe by providing for variations in material properties and loads not anticipated by design calculations. Design factors are based on judgment, past experience, and sound engineering principles.

The design method discussed in this chapter applies in concept to pipe with uniform walls and to pipe with ribbed-wall cross sections. However, for design of pipe with ribbed walls, some of the equations must be modified to allow for the special properties of this pipe. Also, additional calculations not addressed in this chapter may be required to ensure an adequate design for a ribbed-wall cross section. The equations are presented with inch-pound units in the left column and metric units in the right column.

5.2 TERMINOLOGY

5.2.1 Definitions

The following definitions apply to buried pipe design as discussed in this chapter.

Working pressure P_w. The maximum anticipated long-term operating pressure of the fluid system resulting from typical system operation.

Pressure class P_c. The maximum sustained pressure for which the pipe is designed in the absence of other loading conditions.

Surge pressure P_s. The pressure increase above the working pressure, sometimes called *water hammer*, that is anticipated in a system as a result of a change in the velocity of the fluid, such as when valves are operated or when pumps are started or stopped.

Surge allowance P_{sa}. That portion of the surge pressure that can be accommodated without changing pressure class. The surge allowance is expected to accommodate pressure surges usually encountered in typical systems.

Hydrostatic design basis (*HDB*). The long-term hydrostatic hoop strength of a specific fiberglass pipe material as determined by tests and detailed evaluation procedures in accordance with ANSI/AWWA C950, Fiberglass Pressure Pipe, pressure classes subsection on long-term hydrostatic design pressure. The HDB value shall be specific based on the type of joint; for restrained type of joint, the HDB value shall be from a test with free ends.

Design factor (*FS*). A specific number greater than 1 used to reduce a specific mechanical or physical property to establish a design value for use in calculations.

Compacted. Densification of soil using mechanical compaction equipment or using water.

Uncompacted. Placement of soil that is not compacted but is worked with hand tools such as shovels or rods to ensure all voids are filled.

Percent compaction (*PC*). Percent compaction is calculated by dividing the in-place field density of the soil by the laboratory maximum density of the soil and expressing it as a percentage per ASTM D653, Standard Terminology Relating to Soil, Rock, and Contained Fluids:

Laboratory maximum density is determined by compaction test (ASTM D698, Standard Test Methods for Laboratory Compaction Characteristics of Soil Using Standard Effort (12,400 ft-lbf/ft^3 (600 kN-m/m^3)), or AASHTO T 99, Standard Method of Test for Moisture-Density Relations of Soils Using a 2.5-kg (5.5-lb) Rammer and a 305-mm (12-in.) Drop) for cohesive soils and sands and either ASTM D4253, Standard Test Methods for Maximum Index Density and Unit Weight of Soils Using a Vibratory Table, or D7382, Standard Test Methods for Determination of Maximum Dry Unit Weight of Granular Soils Using a Vibrating Hammer, for cohesionless materials (i.e., gravels and crushed rock and alternatively sands).

5.2.2 Symbols

B_d = trench width at pipe springline, in. (mm)
C_n = scalar calibration factor (used in buckling Eq 5-24)
D = mean pipe diameter, in. (mm)
D_f = shape factor per Sec. 5.7.2 (dimensionless)
D_L = deflection lag factor (dimensionless)
E = ring flexural modulus of elasticity, psi (GPa)
E_H = hoop tensile modulus of elasticity, psi (GPa)
EI = stiffness factor per unit length of pipe wall, in.^{2}-lb/in. (m^2-N/m)
F = load per unit length, lb/in. (N/m)
FS = design factor (dimensionless)
$F/\Delta y$ = pipe stiffness, psi (kPa)
H = burial depth to top of pipe, ft (m)

h = burial depth to top of pipe, in. (m)
h_{int} = depth at which load from wheels interacts, in. (m)
h_w = height of water surface above top of pipe, in. (m)
HDB = hydrostatic design basis, psi (MPa) (for stress basis) or in./in. (mm/mm) (for strain basis)
ID = inside diameter, in. (mm)
I = moment of inertia of pipe wall for ring bending, in.^{4}/in. (mm^4/mm)
I_f = impact factor (dimensionless)
K_x = bedding coefficient (dimensionless)
k_υ = modulus correction factor for Poisson's ratio, υ, of the soil (dimensionless)
L_1 = dimension of area of wheel load at pipe crown depth in the direction of travel, in. (m) (see Sec. 5.7.3.6)
L_2 = dimension of area of wheel load at pipe crown depth transverse to the direction of travel, in. (m) (see Sec. 5.7.3.6)
LLDF = live load distribution factor (dimensionless)
M_P = multiple presence factor (dimensionless)
M_s = composite soil constrained modulus, psi (MPa)
M_{sb} = constrained modulus of the embedment, psi (MPa)
M_{sn} = constrained modulus of the native soil at pipe elevation, psi (MPa)
OD = outside diameter, in. (mm)
P = vehicular traffic load (wheel load), lb (N)
P_c = pressure class, psi (kPa)
P_s = surge pressure, psi (kPa)
P_υ = internal vacuum pressure, psi (kPa)
P_w = working pressure, psi (kPa)
PC = percent compaction (see Sec. 5.2.1 and ASTM D653)
PS = pipe stiffness, psi (kPa)
q_u = unconfined compressive strength, US tons/ft^2 (kPa)
r = mean pipe radius, in. (mm)
r_c = rerounding, coefficient (dimensionless)
R_h = buckling strength correction factor for depth of fill (dimensionless)
R_w = water buoyancy factor (dimensionless)
S_b = long-term ring-bending strain, in./in. (mm/mm)
S_c = soil support combining factor (dimensionless)
S_i = ultimate hoop tensile strength, psi (MPa)
S_r = hoop tensile stress, psi (MPa), or strain, in./in. (mm/mm) at pressure class
t = thickness of pipe-reinforced wall, per ASTM D3567, Standard Practice for Determining Dimensions of "Fiberglass" (Glass-Fiber-Reinforced Thermosetting Resin) Pipe and Fittings, in. (mm)
t_L = thickness of liner, in. (mm)
t_l = length of tire footprint, in. (mm)
t_w = width of tire footprint, in. (mm)
t_t = total thickness of pipe wall, in. (mm)
W_c = vertical soil load on the pipe, psi (N/m^2)
W_L = live load on the pipe, psi (N/m^2)
γ_s = unit weight of the soil, lb/ft^3 (N/m$^{3)}$
γ_w = unit weight of water, lb/in.^{3} (N/m^3)
υ = Poisson's ratio of soil (dimensionless)
Δy = predicted vertical pipe deflection, in. (mm)
Δy_a = limiting vertical pipe deflection, in. (mm)
Δy_t = vertical pipe deflection—in. (mm) when tested by ASTM D2412, Standard Test Method for Determination of External Loading Characteristics of Plastic Pipe by Parallel-Plate Loading, with a vertical diameter reduction of 5 percent

σ_b = ring-bending stress due to deflection, psi (MPa)
σ_{pr} = hoop stress due to internal working pressure, psi (MPa)
δ_d = maximum permitted long-term installed deflection (as specified by the engineer or manufacturer), in. (mm)
ϕ_s = factor to account for variability in stiffness of compacted soil (dimensionless)
ε_b = ring-bending strain due to deflection, in./in. (mm/mm)
ε_{pr} = hoop strain due to internal working pressure, in./in. (mm/mm)

5.3 DESIGN CONDITIONS

Design conditions are largely determined by required flow rate and flow velocity limitations, hydraulics, pipeline elevations and associated geology and topography, available right-of-ways, and installation requirements.

5.3.1 Head Losses

Hydraulic head loss due to pipe friction may be significantly lower for fiberglass pipe than for other types of pipe because of fiberglass pipe's generally smoother bores and freedom from tuberculation and corrosion. This is reflected in typical long-term flow coefficient values of 0.009 for Manning's n and 150 for the Hazen–Williams C. The engineer may wish to consider these values when establishing design conditions. (See Chapter 4 on hydraulics.)

5.3.2 Surge Pressures

Surge pressures should be calculated on the basis of the pipe hoop tensile modulus and thickness-to-diameter ratio for given system design parameters (see Chapter 4 and Sec. 5.7.1.3). Excessive surge pressures should be identified in the design phase, and the causative condition should be eliminated, or automatic surge pressure relief provided. Otherwise, a higher pressure class should be selected.

5.3.3 Basic Design Conditions

Design conditions that should be established before performing structural design calculations are as follows:

- Nominal pipe size (Tables 1 through 6 of ANSI/AWWA C950)
- Working pressure, P_w
- Surge pressure, P_s
- Soil conditions for the embedment and native material at pipe depth
- Soil unit weight, γ_s
- Depth of cover, minimum and maximum
- Vehicular traffic load, P
- Internal vacuum pressure, P_v
- Average and maximum service temperatures

5.4 PIPE PROPERTIES

Preliminary pipe pressure class selection can usually be made on the basis of working pressure, surge pressure, and external loads established in Sec. 5.7. Properties at the anticipated average and maximum service temperatures for a given class of a specific pipe product should be obtained from the manufacturer or the manufacturer's literature. Values for ring stiffness, axial strength, and hoop tensile strength given in ANSI/AWWA C950 are minimum requirements. Some pipe products may have significantly higher values for these properties. The design may require material properties and structural capacities greater than those given as minimums in ANSI/AWWA C950. Pipe properties necessary for design calculations include the following:

- Nominal reinforced wall thickness t and liner thickness t_L (ANSI/AWWA C950)
- Hoop tensile modulus of elasticity, E_H
- HDB
- Ring flexural modulus of elasticity, E
- Minimum pipe stiffness, $F/\Delta y$ (ANSI/AWWA C950)
- Long-term ring-bending strain, S_b

5.5 INSTALLATION PARAMETERS

The primary installation parameters that must be selected according to the site conditions and planned installation are the type of embedment, PC, and the characteristics of native soil at the pipe elevation. Initial selection of these parameters may be controlled by prevailing standard specifications, the project soil's boring report, manufacturers' recommendations, or past experience. A given combination of native soil characteristics, embedment material, and PC will largely determine the following values required for design calculations:

- Bedding coefficient, K_x (Sec. 5.7.3.4)
- Constrained modulus of the native soil (M_{sn}) and of the embedment (M_{sb}) (Sec. 5.7.3.8)
- Deflection lag factor, D_L (Sec. 5.7.3.3)

5.6 DESIGN PROCEDURE

With conditions, properties, and installation parameters established in accordance with Sec. 5.3 through Sec. 5.5, satisfaction of the requirements listed in Sec. 5.7 can be checked by design calculations. The calculations may be made using either stress or strain, depending on the basis used to establish a particular product performance limit. The procedure for using design calculations to determine whether pipe meets the requirements discussed in Sec. 5.7 is as follows:

1. Calculate P_c from HDB and pipe dimensions (Sec. 5.7.1.1).
2. Check working pressure, P_w (Sec. 5.7.1.2).
3. Check surge pressure, P_s (Sec. 5.7.1.3).
4. Calculate limiting deflection from ring bending (Sec. 5.7.2).
5. Determine soil loads, W_c, and live loads, W_L (Sec. 5.7.3.5 and Sec. 5.7.3.6, respectively).
6. Calculate the composite soil constrained modulus, M_s (Sec. 5.7.3.8).
7. Check deflection prediction, $\Delta y/D$ (Sec. 5.7.3.2).
8. Check combined loading (Sec. 5.7.4).
9. Check buckling (Sec. 5.7.5).

See Sec. 5.10 for step-by-step example design calculations.

5.7 DESIGN CALCULATIONS AND REQUIREMENTS

5.7.1 Internal Pressure

5.7.1.1 Pressure class P_c. The pressure class in ANSI/AWWA C950 is related to the long-term strength, or HDB, of the pipe as follows:

For stress basis HDB:

$$P_c < \left(\frac{HDB}{FS}\right)\left(\frac{2t}{D}\right) \qquad P_c < \left(\frac{HDB}{FS}\right)\left(\frac{2t}{D}\right) \times 10^3 \qquad \text{(Eq 5-1)}$$

For strain basis HDB:

$$P_c < \left(\frac{HDB}{FS}\right)\left(\frac{2tE_H}{D}\right) \qquad P_c < \left(\frac{HDB}{FS}\right)\left(\frac{2tE_H}{D}\right) \times 10^6 \qquad \text{(Eq 5-2)}$$

Where:

P_c = pressure class, psi
HDB = hydrostatic design basis, psi for stress basis, or in./in. for strain basis
FS = minimum design factor, 1.8
t = pipe-reinforced wall thickness, in.
D = mean pipe diameter, in.
$D = \text{ID} + 2t_L + t$; or
$D = \text{OD} - t$

Where:

t_L = liner thickness, in.
ID = inside diameter, in.
OD = outside diameter, in.
E_H = hoop tensile modulus of elasticity, psi

Where:

P_c = pressure class, kPa
HDB = hydrostatic design basis, MPa, for stress basis, or mm/mm for strain basis
FS = minimum design factor, 1.8
t = pipe-reinforced wall thickness, mm
D = mean pipe diameter, mm
$D = \text{ID} + 2t_L + t$; or
$D = \text{OD} - t$

Where:

t_L = liner thickness, mm
ID = inside diameter, mm
OD = outside diameter, mm
E_H = hoop tensile modulus of elasticity, GPa

HDB. The *HDB* of fiberglass pipe varies for different products, depending on the materials and composition used in the reinforced wall and in the liner. The *HDB* may be defined in terms of reinforced wall hoop stress or hoop strain on the inside surface.

Temperature and service life. The *HDB* at ambient temperature must be established by testing in accordance with ANSI/AWWA C950 for each fiberglass pipe product by each manufacturer. The required practice is to define projected product performance limits at 50 years. Performance limits at elevated temperature depend on the materials and type of pipe wall construction used. The manufacturer should be consulted for *HDB* values appropriate for elevated temperature service.

Design factors. Two separate design factors are required in ANSI/AWWA C950 for internal pressure design.

The first design factor is the ratio of short-term ultimate hoop tensile strength S_i to hoop tensile stress S_r at pressure class P_c. This factor ensures that the stress or strain caused by the short-term peak pressure conditions does not exceed the short-term hydrostatic strength of the pipe. The hoop tensile strength values given in ANSI/AWWA C950's Table 10 reflect a minimum design factor of 4.0 on initial hydrostatic strength.

The second design factor is the ratio of HDB to hoop stress or strain S_r at pressure class P_c. This factor ensures that stress or strain caused by sustained working pressure does not exceed the long-term hoop strength of the pipe as defined by *HDB*. For fiberglass pipe design, this minimum design factor is 1.8.

Both design factors should be checked. Either design factor may govern pipe design, depending on long-term strength regression characteristics of the particular pipe product. Prudent design practice may dictate an increase or decrease in either design factor, depending on the certainty of the known service conditions.

5.7.1.2 Working pressure P_w. The pressure class of the pipe should be equal to or greater than the working pressure in the system, as follows:

$$P_c \geq P_w \qquad\qquad P_c \geq P_w \qquad \text{(Eq 5-3)}$$

Where:

P_w = working pressure, psi $\qquad$ P_w = working pressure, kPa

5.7.1.3 Surge pressure P_s. The pressure class of the pipe should be equal to or greater than the maximum pressure in the system—working pressure plus surge pressure—divided by 1.4, as follows:

$$P_c \geq \frac{(P_w + P_s)}{1.4} \qquad\qquad P_c \geq \frac{(P_w + P_s)}{1.4} \qquad \text{(Eq 5-4)}$$

Where:

P_s = surge pressure, psi $\qquad$ P_s = surge pressure, kPa

The treatment of surge pressures reflects the characteristics of the pipe and materials covered by ANSI/AWWA C950. Factory hydrotesting at pressures up to 2 P_c is acceptable and is not governed by Eqs 5-3 and 5-4.

Calculated surge pressure P_s. The surge-pressure calculations should be performed using recognized and accepted theories. (See Chapter 4 on hydraulics.)

Calculated surge-pressure magnitudes are highly dependent on the hoop tensile elastic modulus and thickness-to-diameter (*t*/*D*) ratio of the pipe. Because of this, the engineer should generally expect lower calculated surge pressures for fiberglass pipe than for pipe materials with a higher modulus or thicker wall or both. For example, an instantaneous change in flow velocity of 2 ft/s (0.6 m/s) would result in a calculated surge-pressure increase of approximately 40 psi (276 kPa) for fiberglass pipe with a modulus of 3,000,000 psi (20.7 GPa) and a *t*/*D* ratio of 0.01.

Surge allowance. The surge allowance is intended to provide for rapid transient pressure increases typically encountered in transmission systems. The surge-pressure allowance of 0.4 P_c is based on the increased strength of fiberglass pipe for rapid strain rates. Special consideration should be given to the design of systems subject to rapid and frequent cyclic service. The manufacturer should be consulted for specific recommendations.

5.7.2 Ring Bending

The long-term vertical pipe deflection should not exceed the limiting deflection, which is the vertical pipe deflection corresponding to the long-term ring-bending strain (or stress) capacity of the pipe reduced by an appropriate design factor. Satisfaction of this requirement is assured by using one of the following formulas:

For stress basis:

$$\sigma_b = D_f E\left(\frac{\Delta y_a}{D}\right)\left(\frac{t_t}{D}\right) \leq \frac{S_b E}{FS} \qquad\qquad \sigma_b = 10^3 D_f E\left(\frac{\Delta y_a}{D}\right)\left(\frac{t_t}{D}\right) \leq 10^3\frac{S_b E}{FS} \qquad \text{(Eq 5-5)}$$

For strain basis:

$$\varepsilon_b = D_f\left(\frac{\Delta y_a}{D}\right)\left(\frac{t_t}{D}\right) \leq \frac{S_b}{FS} \qquad \varepsilon_b = D_f\left(\frac{\Delta y_a}{D}\right)\left(\frac{t_t}{D}\right) \leq \frac{S_b}{FS} \qquad \text{(Eq 5-6)}$$

Where:

σ_b = ring-bending stress due to deflection, psi
D_f = shape factor per Table 5-1, dimensionless
E = ring flexural modulus of elasticity for the pipe, psi
Δy_a = limiting vertical pipe deflection, in.
S_b = long-term ring-bending strain for the pipe, in./in.
D = mean pipe diameter, in.
FS = design factor, 1.5
ε_b = ring-bending strain due to deflection, in./in.
t_t = total wall thickness, in.
t_t = $t + t_L$

Where:

σ_b = ring-bending stress due to deflection, MPa
D_f = shape factor per Table 5-1, dimensionless
E = ring flexural modulus of elasticity for the pipe, GPa
Δy_a = limiting vertical pipe deflection, mm
S_b = long-term ring-bending strain for the pipe, mm/mm
D = mean pipe diameter, mm
FS = design factor, 1.5
ε_b = ring-bending strain due to deflection, mm/mm
t_t = total wall thickness, mm
t_t = $t + t_L$

5.7.2.1 Shape factor D_f. The shape factor relates pipe deflection to bending stress or strain and is a function of pipe stiffness, embedment material and PC, haunching, native soil conditions, and level of deflection. Table 5-1 gives values for D_f, assuming inconsistent haunching, deflections of at least 2–3 percent, and stable native soils or adjustments to trench width to offset poor conditions. Values given in Table 5-1 are for typical embedment materials. For finer-grain-size embedment materials (Class IV), the D_f value for soil with PC greater than or equal to 85 percent (ASTM D698) should be used.

5.7.2.2 Long-term ring-bending strain S_b. The long-term ring-bending strain varies for different products, depending on materials and type of construction used in the pipe wall. Long-term ring-bending strain should be determined as defined in ANSI/AWWA C950.

5.7.2.3 Bending design factor. Prudent design of pipe to withstand bending requires consideration of two separate design factors.

The first design consideration is comparison of initial deflection at failure to the maximum allowed installed deflection. The ring stiffness test (level B) in ANSI/AWWA C950 subjects a pipe ring to deflections far exceeding those permitted in use. This test requirement demonstrates a design factor of at least 2.5 on initial bending strain.

Table 5-1 Shape factors D_f

		Embedment Material and Percent Compaction*			
		Gravel†		Sand‡	
Pipe Stiffness		<85%	≥85%	<85%	≥85%
psi	*kPa*	Shape Factor D_f (Dimensionless)			
18	124	4.5	5.5	5.0	6.5
36	248	3.8	4.5	4.0	5.5
46	317	3.7	4.3	3.9	5.2
72	496	3.3	3.8	3.5	4.5

* Percent compaction based on maximum density determined by ASTM D698, ASTM D4253, or ASTM D7382, as applicable.

† GW, GP, GW–GC, GW–GM, GP–GC, and GP–GM per ASTM D2487, Standard Classification of Soils for Engineering Purposes (Unified Soil Classification System) (includes crushed rock). For crushed rock and clean gravels, values for D_f shall be based on the ≥85 percent gravel column if compacted and on the <85 percent gravel column if uncompacted.

‡ SW, SP, SM, SC, GM, and GC or mixtures per ASTM D2487.

The second design factor is the ratio of long-term bending stress or strain to the bending stress or strain at the limiting vertical pipe deflection. For fiberglass pipe design, this minimum design factor is 1.5.

5.7.3 Deflection

Buried pipe should be installed in a manner that will ensure that external loads will not cause a long-term decrease in the vertical diameter of the pipe, resulting in exceeding the limiting deflection ($\Delta y_a/D$) established in Sec. 5.7.2 or the permitted deflection ($\delta d/D$) as required by the engineer or the manufacturer, whichever is less. This requirement may be stated as follows:

$$\Delta y / D \leq \delta d / D \leq \Delta y_a / D$$

$$\Delta y / D \leq \delta d / D \leq \Delta y_a / D \qquad \text{(Eq 5-7)}$$

Where:

$\Delta y/D$ = predicted vertical pipe deflection, fraction of mean diameter
$\delta d/D$ = permitted vertical pipe deflection, fraction of mean diameter
$\Delta y_a/D$ = limiting vertical pipe deflection, fraction of mean diameter

Where:

$\Delta y/D$ = predicted vertical pipe deflection, fraction of mean diameter
$\delta d/D$ = permitted vertical pipe deflection, fraction of mean diameter
$\Delta y_a/D$ = limiting vertical pipe deflection, fraction of mean diameter

$$\frac{\Delta y}{D} = \frac{(D_L W_c + W_L) K_x}{0.149 PS + 0.061 M_s}$$

$$\frac{\Delta y}{D} = \frac{(D_L W_c + W_L) K_x}{149 PS + 61{,}000 M_s} \qquad \text{(Eq. 5-8)}$$

Where:

D_L = deflection lag factor, dimensionless
W_c = vertical soil load on pipe, psi
W_L = live load on pipe, psi
K_x = bedding coefficient, dimensionless
PS = pipe stiffness, psi
M_s = composite soil constrained modulus, psi

Where:

D_L = deflection lag factor, dimensionless
W_c = vertical soil load on pipe, N/m^2
W_L = live load on pipe, N/m^2
K_x = bedding coefficient, dimensionless
PS = pipe stiffness, kPa
M_s = composite soil constrained modulus, MPa

NOTE: The deflection calculated by Eq 5-8 is the predicted long-term pipe deflection when the load term (W_c) is the projected long-term vertical soil load on the pipe. This is typically taken as the vertical soil column over the pipe (prism load), which is the maximum soil load for a flexible pipe (see Sec. 5.7.3.5). When calculating the long-term deflection using the prism load, the deflection lag factor (D_L) should be taken as 1.0. To calculate the predicted initial pipe deflection, use the initial vertical soil load for W_c. A procedure to determine the initial vertical soil load may be found in the *Concrete Pipe Design Manual* published by the American Concrete Pipe Association.

5.7.3.1 Deflection calculations. Design calculations that require deflection as an input parameter should show the predicted deflection $\Delta y/D$, as well as the limiting deflection ($\Delta ya/D$) at which the allowable design stress or strain is not exceeded. The maximum permitted deflection $\delta d/D$ (as specified by the engineer or manufacturer) should be used in all design calculations.

5.7.3.2 Deflection prediction. When installed in the ground, all flexible pipe will undergo deflection, defined here to mean a decrease in vertical diameter. The amount of deflection is a function of the soil load, live load, native soil characteristics at pipe elevation, pipe embedment material and density, trench width, haunching, and pipe stiffness. Many theories have been proposed to predict deflection levels; however, in actual field conditions, pipe deflections may vary from calculated values because the actual installation achieved may vary from the installation planned. These variations include the inherent variability of native ground conditions and variations in methods, materials, and equipment used to install a buried pipe.

Field personnel responsible for pipe installation must follow procedures designed to ensure that the long-term pipe deflection is less than the limiting deflection Δy_a as determined in Sec. 5.7.2, or as required by the engineer or manufacturer, whichever is less. As presented previously and as augmented by information provided in the following sections, Eq 5-8 serves as a guideline for estimating the expected level of short-term and long-term deflection that can be anticipated in the field. Eq 5-8 is a form of the Iowa formula, first published by M.G. Spangler (1941). This equation is the best known and documented of a multitude of deflection-prediction equations that have been proposed. As presented in this chapter, the Iowa formula treats the major aspects of pipe-soil interaction with sufficient accuracy to produce reasonable estimates of load induced field deflection levels.

Pipe deflection due to self-weight and initial ovalization caused by pipe embedment placement and compaction are not addressed by this method. These deflections are typically small for pipe stiffness greater than 18 psi (124 kPa), depending on installation conditions. For pipe stiffness less than these values, consideration of these items may be required to achieve an accurate deflection prediction.

Application of this method is based on the assumption that the design values used for bedding, embedment, and compaction levels will be achieved with good practices and appropriate equipment in the field. Experience has shown that deflection levels of any flexible conduit can be higher or lower than predicted by calculation if the design assumptions are not achieved.

5.7.3.3 Deflection lag factor D_L. Traditionally, the deflection lag factor (D_L) is the ratio of the final long-term pipe deflection to the initial pipe deflection at the time of installation. Although this definition varies from Spangler's original recommendation, in current practice, the lag factor is generally used to account for increased deflection from any cause, including increases in soil load and consolidation of the embedment or the native soil adjacent to the pipe. Increases in pipe deflection with time typically occur during the first few weeks or months after installation but may continue for some years, depending on the frequency of wetting and drying cycles, surface loads, and the amount of original compaction of the backfill.

However, a distinction between field observations and calculations in this manual must be noted. In practice, pipe deflection generally increases with time. This is mostly because of increasing overburden load over time to reach the maximum load for a flexible pipe, which is the weight of the soil column directly above. As explained in the previous note, the load term (W_c) in Eq 5-8 is the projected long-term vertical soil load on the pipe, so the factor, D_L, to account for load increasing from initial to long-term, must be taken near 1.0 in Eq 5-8 to accurately compute the projected long-term deflection. If the initial overburden soil load is used for *Wc*, a D_L greater than 1.0 would be needed to compute the long-term deflection. For shallowly buried pipes in a high-quality installation, the long-term load increase is small, so D_L is not much greater than 1.0. For very deeply buried pipe, the load increase over time may be large; therefore, D_L may approach 2.0 or greater.

5.7.3.4 Bedding coefficient K_x. The bedding coefficient reflects the degree of support provided by the embedment at the bottom of the pipe and over which the bottom reaction is distributed. Assuming an inconsistent haunch achievement (typical direct bury condition), a K_x value of 0.1 should be used. For uniform shaped bottom support, a K_x value of 0.083 is appropriate.

5.7.3.5 Vertical soil load on the pipe W_c. The long-term vertical soil load on the pipe may be considered as the weight of the rectangular prism of soil directly above the pipe. The soil prism would have a height equal to the depth of earth cover and a width equal to the pipe outside diameter (OD).

$$W_c = \frac{\gamma_s H}{144} \qquad W_c = \gamma_s H \qquad \text{(Eq 5-9)}$$

Where:
W_c = vertical soil load, psi
γ_s = unit weight of soil overburden, lb/ft^3
H = burial depth to top of pipe, ft

Where:
W_c = vertical soil load, N/m^2
γ_s = unit weight of soil overburden, N/m^3
H = burial depth to top of pipe, m

5.7.3.6 Live loads on the pipe, W_L. The following calculations may be used to compute the live load on the pipe for surface traffic. The procedure is based on the requirements of the American Association of State Highway and Transportation Officials Load-and-Resistance Factor Design philosophy from *LRFD Bridge Design Specifications* (AASHTO 2024). These calculations consider a single-axle truck traveling perpendicular to the pipe on an unpaved surface or a road with flexible pavement. With the inclusion of the multiple presence factor (M_P), the previous conditions generally control and may be assumed to yield acceptably conservative load estimates. The multiple presence factor (M_P) accounts for the probability relationship between the number of loaded lanes and the weight as any single vehicle. M_P is typically taken as 1.2 for single passing vehicle (see Figure 5-1).

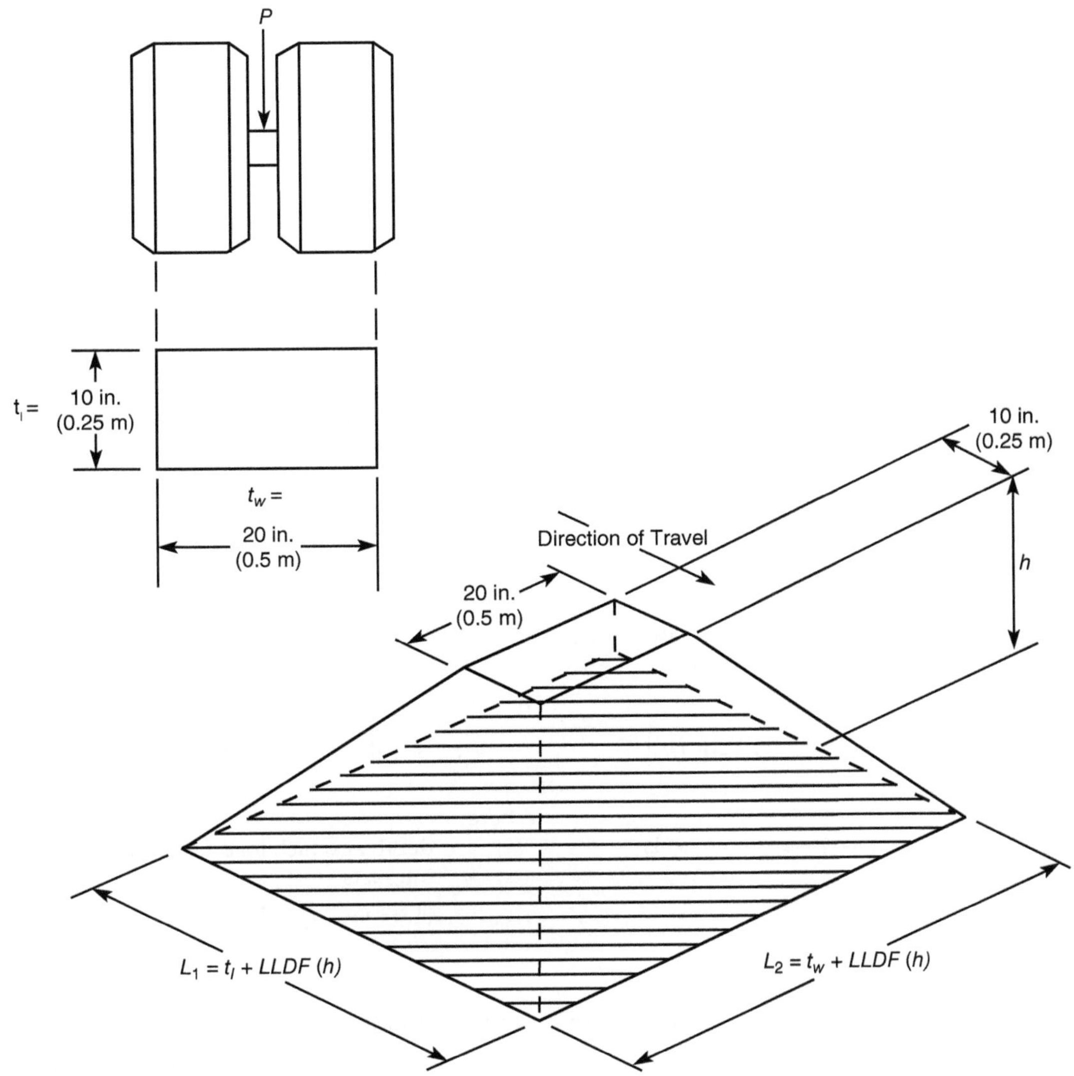

NOTE: For $h > 45$ in. (1.14 m), see part 2 of the L_2 formula. (Change accounts for overlapping influence areas from adjacent wheel loads.)

Figure 5-1 Distribution of AASHTO HS-20 or HS-25 live load through granular fill for h ≤ 45 in. (1.14 m)

$$W_L = \frac{M_p P I_f}{(L_1)(L_2)}$$

Where:

W_L = live load on pipe, psi
M_P = multiple presence factor = 1.2
P = wheel load
= 16,000 lb for AASHTO HS-20 truck
= 20,000 lb for AASHTO HS-25 truck
I_f = impact factor
L_1 = load width parallel to direction of travel, in.
L_2 = load width perpendicular to direction of travel, in.

$$W_L = \frac{M_p P I_f}{(L_1)(L_2)} \quad \text{(Eq 5-10)}$$

Where:

W_L = live load on pipe, N/m²
M_P = multiple presence factor = 1.2
P = wheel load
= 71,300 N for AASHTO HS-20 truck
= 89,000 N for AASHTO HS-25 truck
I_f = impact factor
L_1 = load width parallel to direction of travel, m
L_2 = load width perpendicular to direction of travel, m

$$I_f = 1 + 0.33\left[(96 - h)/96\right] \geq 1.0$$

Where:

h = burial depth to top of pipe, in.

$$I_f = 1 + 0.33\left[(2.44 - h)/2.44\right] \geq 1.0 \quad \text{(Eq 5-11)}$$

Where:

h = burial depth to top of pipe, m

$$L_1 = t_l + \text{LLDF}\,(h)$$

Where:

t_l = length of tire footprint = 10 in.
LLDF = factor to account for live load distribution with depth of fill
= 1.15 for backfills Class I and Class II (Sec. 6.5)
= 1.0 for all other backfills

$$L_1 = t_l + \text{LLDF}\,(h) \quad \text{(Eq 5-12)}$$

Where:

t_l = length of tire footprint = 0.25 m
LLDF = factor to account for live load distribution with depth of fill
= 1.15 for backfills Class I and Class II (Sec. 6.5)
= 1.0 for all other backfills

If $h \leq h_{int}$

$$L_2 = t_w + \text{LLDF}\,(h)$$

Where:

t_w = width of tire footprint = 20 in.

If $h \leq h_{int}$

$$L_2 = t_w + \text{LLDF}\,(h) \quad \text{(Eq 5-13)}$$

Where:

t_w = width of tire footprint = 0.5 m

If $h > h_{int}$

$$L_2 = \left[t_w + 72\ in. + \text{LLDF}\,(h)\right]/2$$

Where:

h_{int} = depth at which load from wheels interacts

$$h_{int} = (72\ in. - t_w)/\text{LLDF}$$

If $h > h_{int}$

$$L_2 = \left[t_w + 1.83\ m + \text{LLDF}\,(h)\right]/2 \quad \text{(Eq 5-14)}$$

Where:

h_{int} = depth at which load from wheels interacts

$$h_{int} = (1.83\ m - t_w)/\text{LLDF} \quad \text{(Eq 5-15)}$$

Calculation notes follow:

1. Equations as shown are for h in inches (meters).
2. AASHTO also specifies a surface lane load of 640 lb/ft (9,350 N/m) over a 10-ft (3-m) lane width. This lane load is ignored in these calculations because it has only a small effect on the total live load and may be added by the engineer if deemed appropriate.

3. The above calculation method assumes that the live load extends over the full diameter of the pipe. This may be conservative for large-diameter pipe under low fills. To account for this, the calculated live load pressure on the pipe may be reduced by the ratio L_1 /OD if the truck is moving across the pipe and $L_1 <$ OD or by L_2/OD if the truck is moving parallel to the pipe and $L_2 <$ OD.
4. For depths of fill less than 2 ft (0.6 m) or for live load magnitudes greater than HS-25, it may be necessary to consider the local live load effects at the crown of the pipe. Such an analysis is beyond the scope of this manual.
5. The previous calculation is for single-axle trucks. Design for tandem-axle trucks may use the same procedures; however, the following substitutions for L_1 should be used if both axles load the pipe at the same time.

$$L_1 = \left[\text{axle spacing} + t_l + \text{LLDF}\ (h)\right]/\,2 \qquad L_1 = \left[\text{axle spacing} + t_l + \text{LLDF}\ (h)\right]/\,2 \qquad \text{(Eq 5-16)}$$

Tandem-axle wheel loads are usually lighter than HS-20 or HS-25 trucks. For example, the AASHTO LRFD design tandem has a 12,500-lb (55,700-N) wheel load.

6. Rigid pavements dramatically reduce live load effects on concrete pipe. The Portland Cement Association developed a calculation method to consider loads transmitted through concrete pavements (Portland Cement Association 1950) that is still currently in use and is suitable for computing live loads on fiberglass pipe under rigid pavements. The same method is also presented in the *Concrete Pipe Design Manual*.

Table 5-2 presents computed live loads for AASHTO HS-20 and HS-25 single-axle trucks based on an LLDF of 1.15 (for Class I and Class II backfill—see Sec. 6.5). The loads shown assume that the load extends over the full diameter of the pipe. This assumption will not be true for large-diameter pipes with shallow covers. Loads for this condition may be lower. See calculation note 3 for guidance on appropriate adjustments.

Table 5-2 AASHTO HS-20, HS-25, and Cooper E80 live loads

AASHTO Live Loads		HS-20		HS-25		Coopers E80 Live Loads*			
Depth		W_L		W_L		Depth		W_L	
ft	*m*	*psi*	*kPa*	*psi*	*kPa*	*ft*	*m*	*psi*	*kPa*
2	0.6	13.4	92	16.8	116	3	0.9	16	110
2.5	0.8	9.7	67	12.2	84	4	1.2	14.1	97
3	0.9	7.4	51	9.2	63	5	1.5	12.2	84
4	1.2	4.7	32	5.9	41	6	1.8	10.5	72
5	1.5	3.4	23	4.2	29	7	2.1	9	62
6	1.8	2.6	18	3.2	22	8	2.4	7.7	53
8	2.4	1.6	11	2	14	10	3.0	5.7	39
10	3.0	1.1	7.6	1.4	10	12	3.7	4.6	32
12	3.7	0.8	5.5	1.1	7.6	15	4.6	3.4	23
15	4.6	0.6	4.1	0.7	4.8	20	6.1	2.2	15
20	6.1	0.4	2.8	0.5	3.4	25	7.6	1.5	10
28	8.5	0.2	1.4	0.25	1.8	30	9.1	1.1	7.6
40	12.2	0.1	0.7	0.1	0.7	40	12.2	0.6	4.1

* Cooper E80 as defined by American Railway Engineers and Maintenance-of-Way Association (AREMA) in its *Manual of Railway Engineering*.
NOTE: Live loads presented are for single-axle loads and do not include effects of axle interaction.

Also included in Table 5-2 are live loads from Cooper E80 railroad loading (taken from the *Concrete Pipe Design Manual*).

Figures 5-2 through 5-4 graphically depict the AASHTO truck and Cooper E80 railroad live loads shown in Table 5-2.

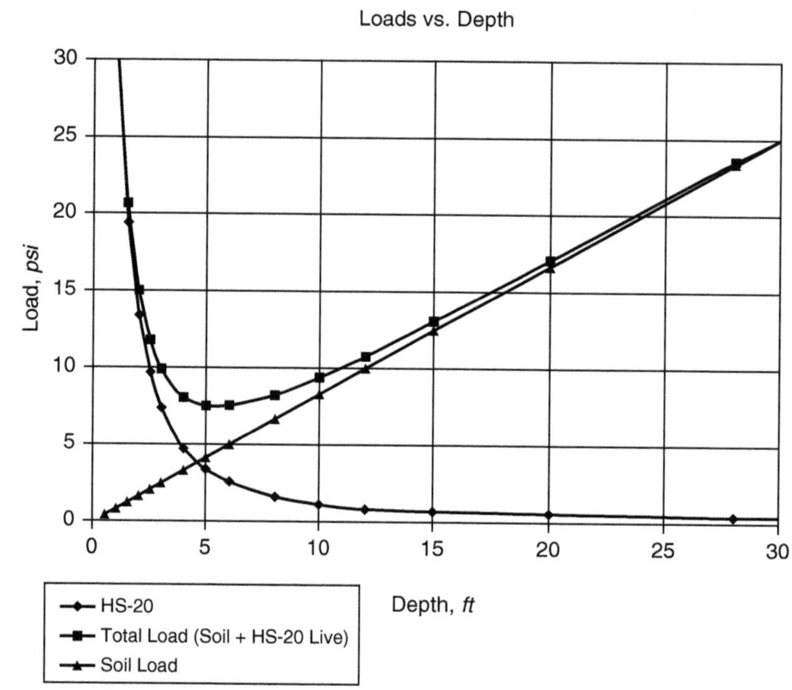

Figure 5-2 AASHTO HS-20 live load, soil load (120 lb/ft³), and total load graph

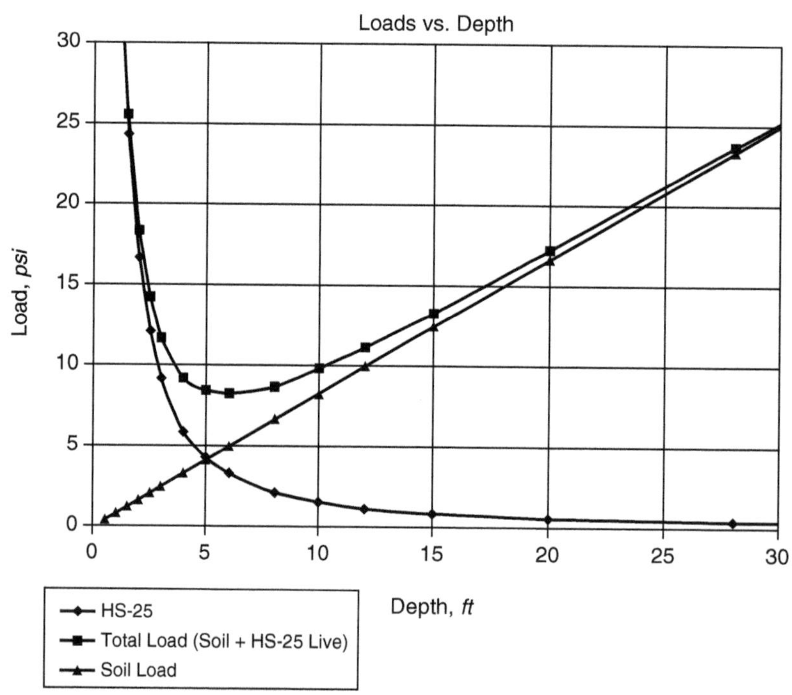

Figure 5-3 AASHTO HS-25 live load, soil load (120 lb/ft³), and total load graph

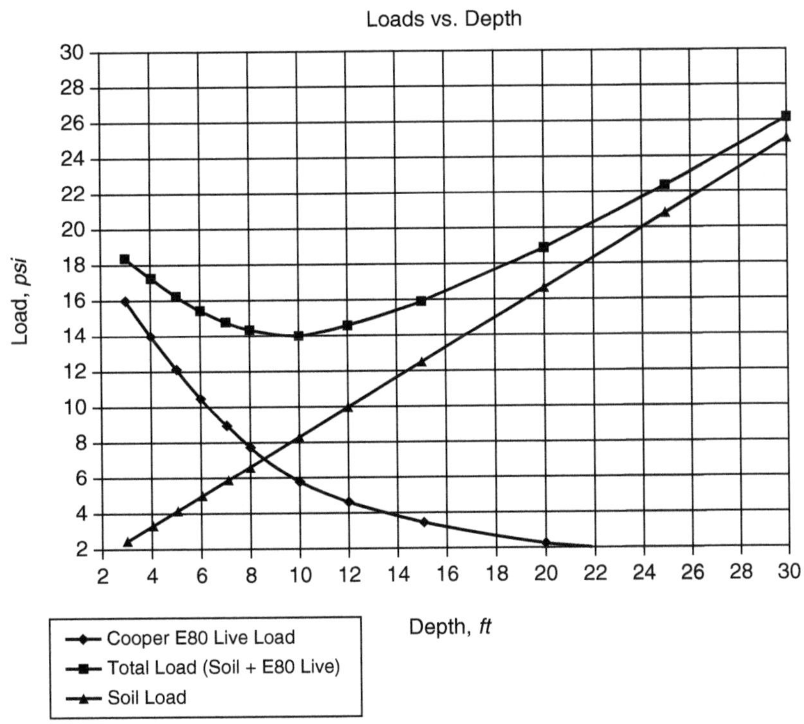

Figure 5-4 Cooper E80 live load, soil load (120 lb/ft³), and total load graph

5.7.3.7 Pipe stiffness, PS. The pipe stiffness can be determined by conducting parallel-plate loading tests in accordance with ASTM D2412. During the parallel-plate loading test, deflection caused by loads on the top and bottom of the pipe is measured, and pipe stiffness is calculated from the following equation:

$$PS = \frac{F}{\Delta y_t} \qquad PS = \frac{1{,}000F}{\Delta y_t} \qquad \text{(Eq. 5-17)}$$

Where:
- PS = pipe stiffness, psi
- F = load per unit length, lb/in.
- Δy_t = vertical pipe deflection, in., when tested by ASTM D2412 with a vertical diameter reduction of 5%

Where:
- PS = pipe stiffness, kPa
- F = load per unit length, N/mm
- Δy_t = vertical pipe deflection, mm, when tested by ASTM D2412 with a vertical diameter reduction of 5%

Pipe stiffness may also be determined by the pipe dimensions and material properties using Eq 5-18:

$$PS = \frac{EI}{0.149 \times r^3 \times C} \qquad PS = \frac{EI \times 10^6}{0.149 \times r^3 \times C} \qquad \text{(Eq 5-18)}$$

Where:

$$C = \left(1 + \frac{\Delta y_t}{2 \times ID}\right)^3$$

- E = ring flexural modulus, psi
- I = moment of inertia of unit length, in.⁴/in.
  = $(t_t)^3/12$
- r = mean pipe radius, in.
  = $(OD - t)/2$

Where:

$$C = \left(1 + \frac{\Delta y_t}{2 \times ID}\right)^3$$

- E = ring flexural modulus, GPa
- I = moment of inertia of unit length, mm⁴/mm
  = $(t_t)^3/12$
- r = mean pipe radius, mm
  = $(OD - t)/2$

5.7.3.8 Composite soil constrained modulus, M_s. The vertical loads on a flexible pipe cause a decrease in the vertical diameter and an increase in the horizontal diameter. The horizontal movement develops a passive soil resistance that helps support the pipe. The passive soil resistance varies depending on the soil type and the PC of the embedment material; native soil characteristics at pipe elevation, cover depth, and trench width (see Table 5-3).

The historical parameter used to characterize the soil stiffness in design of flexible pipe is the modulus of soil reaction, E'. In 2000, AASHTO adopted new soil stiffness values for embedment materials around thermoplastic pipe, including changing the soil design parameter from E' to the constrained modulus, M_s. This change is based on the work of McGrath (1998). Design values of the constrained modulus are presented in Table 5-4. The table shows that M_{sb} increases with depth of fill, which reflects the increased confining pressure. This is a well-known soil behavior. At moderate depths of fill, the values of M_{sb} are close to the E' values proposed by Howard (1977, 1996). In design for deflection control, M_s may be substituted directly for E' in the Iowa formula.

To determine M_s for a buried pipe, separate M_s values for the native soil, M_{sn}, and the embedment, M_{sb}, must be determined and then combined using Eq 5-19. Special cases are discussed later in this chapter.

$$M_s = S_c M_{sb}$$

Where:

M_s = composite soil constrained modulus, psi
S_c = soil support combining factor from Table 5-5, dimensionless
M_{sb} = constrained modulus of the embedment from Table 5-4, psi

To use Table 5-5 for S_c, the following values are required:

M_{sn} = constrained modulus of the native soil at pipe elevation, psi (see Table 5-6)
B_d = trench width at pipe springline, in.

$$M_s = S_c M_{sb} \qquad \text{(Eq 5-19)}$$

Where:

M_s = composite soil constrained modulus, MPa
S_c = soil support combining factor from Table 5-5, dimensionless
M_{sb} = constrained modulus of the embedment from Table 5-4, MPa

To use Table 5-5 for S_c, the following values are required:

M_{sn} = constrained modulus of the native soil at pipe elevation, MPa (see Table 5-6)
B_d = trench width at pipe springline, mm

Table 5-3 Soil classification chart

Criteria for Assigning Group Symbols and Group Names Using Laboratory Tests[a]				Soil Classification: Group Symbol	Soil Classification: Group Name[b]
Coarse-grained soils More than 50% retained on No. 200 sieve	Gravels More than 50% of coarse fraction retained on No. 4 sieve	Clean gravels Less than 5% fines[c]	$C_u \geq 4$ and $1 \leq C_c \leq 3$[e]	GW	Well-graded gravel[f]
			$C_u < 4$ and/or $1 > C_c > 3$[e]	GP	Poorly graded gravel[f]
		Gravels with fines More than 12% fines[c]	Fines classify as ML or MH	GM	Silty gravel[f,g,h]
			Fines classify as CL or CH	GC	Clayey gravel[f,g,h]
	Sands 50% or more of coarse fraction passes No. 4 sieve	Clean sands Less than 5% fines[d]	$C_u \geq 6$ and $1 \leq C_c \leq 3$[e]	SW	Well-graded sand[i]
			$C_u < 6$ and/or $1 > C_c$ 3[e]	SP	Poorly graded sand[i]
		Sands with fines More than 12% fines[d]	Fines classify as ML or MH	SM	Silty sand[g,h,i]
			Fines classify as CL or CH	SC	Clayey sand[g,h,i]
Fine-grained soils 50% or more passes the No. 200 sieve	Silts and clays Liquid limit less than 50	Inorganic	PI > 7 and plots on or above "A" line[j]	CL	Lean clay[k,l,m]
			PI < 4 or plots below "A" line[j]	ML	Silt[k,l,m]
		Organic	(Liquid limit—oven dried) / (Liquid limit—not dried) < 0.75	OL	Organic clay[k,l,m,n]
					Organic silt[k,l,m,o]
	Silts and clays Liquid limit 50 or more	Inorganic	PI plots on or above "A" line	CH	Fat clay[k,l,m]
			PI plots below "A" line	MH	Elastic silt[k,l,m]
		Organic	(Liquid limit—oven dried) / (Liquid limit—not dried) < 0.75	OH	Organic clay[k,l,m,p]
					Organic silt[k,l,m,q]
Highly organic soils		Primarily organic matter, dark in color, and organic odor		PT	Peat

a Based on the material passing the 3-in. (75-mm) sieve.
b If field sample contained cobbles and/or boulders, add "with cobbles and/or boulders" to group name.
c Gravels with 5% to 12% fines require dual symbols:
GW–GM well-graded gravel with silt
GW–GC well-graded gravel with clay
GP–GM poorly graded gravel with silt
GP–GC poorly graded gravel with clay
d Sands with 5% to 12% fines require dual symbols:
SW–SM well-graded sand with silt
SW–SC well-graded sand with clay
SP–SM poorly graded sand with silt
SP–SC poorly graded sand with clay
e

$$C_u = D_{60}/D_{10}$$

$$C_c = \frac{(D_{30})^2}{D_{10} \times D_{60}}$$

f If soil contains ≥ 15% sand, add "with sand" to group name.
g If fines classify as CL–ML, use dual symbol GC–GM or SC–SM.
h If fines are organic, add "with organic fines" to group name.
i If soil contains ≥ 15% gravel, add "with gravel" to group name.
j If the Atterberg limits (liquid limit and plasticity index) plot in hatched area on plasticity chart, soil is a CL–ML, silty clay.
k If soil contains 15% to 29% plus No. 200, add "with sand" or "with gravel," whichever is predominant.
l If soil contains ≥ 30% plus No. 200, predominantly sand, add "sandy" to group name.
m If soil contains ≥ 30% plus No. 200, predominantly gravel, add "gravelly" to group name.
n PI ≥ 4 and plots on or above "A" line.
o PI ≤ 4 or plots below "A" line.
p PI plots on or above "A" line.
q PI plots below "A" line.

Source: ASTM D2487; reprinted with permission from the *Annual Book of ASTM Standards*, 100 Barr Harbor Dr., West Conshohocken, PA 19428-2959.

NOTE: ASTM D2487 allows the use of "borderline" symbols when test results indicate that the soil classification is close to another group. Borderline condition is indicated by a slash between the two symbols, e.g., CL/CH.

Table 5-4 M_{sb} based on embedment type and percent compaction

Inch-Pound Units				
Vertical Stress Level (see note 5) *psi*	Depth for γ_s = 120 lb/ft^3 *ft*	Soil Class I		
		Compacted* *psi*	Uncompacted *psi*	
1	1.2	2,350	2,000	
5	6	3,450	2,600	
10	12	4,200	3,000	
20	24	5,500	3,450	
40	48	7,500	4,250	
60	72	9,300	5,000	
		Soil Class II		
		95% *psi*	90% *psi*	85% *psi*
1	1.2	2,000	1,275	470
5	6	2,600	1,500	520
10	12	3,000	1,625	570
20	24	3,450	1,800	650
40	48	4,250	2,100	825
60	72	5,000	2,500	1,000
		Soil Class III		
		95% *psi*	90% *psi*	85% *psi*
1	1.2	1,415	670	360
5	6	1,670	740	390
10	12	1,770	750	400
20	24	1,880	790	430
40	48	2,090	900	510
60	72	2,300	1,025	600
		Soil Class IV		
		95% *psi*	90% *psi*	85% *psi*
1	1.2	530	255	130
5	6	625	320	175
10	12	690	355	200
20	24	740	395	230
40	48	815	460	285
60	72	895	525	345

(continued)

Table 5-4 M_{sb} based on embedment type and percent compaction (*Continued*)

Metric Units

Vertical Stress Level (see note 5)	Depth for Soil Density = 18.8 kN/m^3	Soil Class I		
		Compacted*	Uncompacted	
kPa	*m*	*MPa*	*MPa*	
6.9	0.4	16.2	13.8	
34.5	1.8	23.8	17.9	
69	3.7	29	20.7	
138	7.3	37.9	23.8	
276	14.6	51.7	29.3	
414	22	64.1	34.5	
		Soil Class II		
		95% *MPa*	90% *MPa*	85% *MPa*
6.9	0.4	13.8	8.8	3.2
34.5	1.8	17.9	10.3	3.6
69	3.7	20.7	11.2	3.9
138	7.3	23.8	12.4	4.5
276	14.6	29.3	14.5	5.7
414	22	34.5	17.2	6.9
		Soil Class III		
		95% *MPa*	90% *MPa*	85% *MPa*
6.9	0.4	9.8	4.6	2.5
34.5	1.8	11.5	5.1	2.7
69	3.7	12.2	5.2	2.8
138	7.3	13	5.4	3
276	14.6	14.4	6.2	3.5
414	22	15.9	7.1	4.1
		Soil Class IV		
		95% *MPa*	90% *MPa*	85% *MPa*
6.9	0.4	3.7	1.8	0.9
34.5	1.8	4.3	2.2	1.2
69	3.7	4.8	2.5	1.4
138	7.3	5.1	2.7	1.6
276	14.6	5.6	3.2	2
414	22	6.2	3.6	2.4

* For Class I material, "Compacted" implies placement in maximum 9-in. lifts and mechanical compaction of each lift with at least two passes of suitable compaction equipment.

NOTES:

1. Soil classifications are explained in Chapter 6.
2. Vertical stress level is the vertical effective soil stress at the springline elevation of the pipe. It is normally computed as the design soil unit weight times the depth of fill. Buoyant unit weight should be used for the calculation of M_{sb} below the groundwater level. See Eq 5-25 for calculation of water buoyancy factor. (NOTE: For prediction of deflections, W_c is based on dry weight and not corrected for buoyancy.)
3. Engineers may interpolate intermediate values of M_{sb} for vertical stress levels not shown on the table.
4. For pipe installed below the water table, the calculated modulus should be corrected for reduced vertical stress due to buoyancy (see note 2) and then multiplied by a factor of 1.00 for Class I and Class II soils with compaction of ≥95%, 0.85 for Class II soils with compaction of 90%, 0.70 for Class II soils with compaction of 85%, 0.50 for Class III soils, and 0.30 for Class IV soils.
5. It is recommended to embed pipe with stiffness of ≤9 psi (≤62 kPa) only in Class I or Class II soils.

Table 5-5 Values for the soil support combining factor S_c

M_{sn}/M_{sb}	$B_d/D = 1.25$	$B_d/D = 1.5$	$B_d/D = 1.75$	$B_d/D = 2$	$B_d/D = 2.5$	$B_d/D = 3$	$B_d/D = 4$	$B_d/D = 5$
0.005	0.02	0.05	0.08	0.12	0.23	0.43	0.72	1.00
0.01	0.03	0.07	0.11	0.15	0.27	0.47	0.74	1.00
0.02	0.05	0.10	0.15	0.20	0.32	0.52	0.77	1.00
0.05	0.10	0.15	0.20	0.27	0.38	0.58	0.80	1.00
0.1	0.15	0.20	0.27	0.35	0.46	0.65	0.84	1.00
0.2	0.25	0.30	0.38	0.47	0.58	0.75	0.88	1.00
0.4	0.45	0.50	0.56	0.64	0.75	0.85	0.93	1.00
0.6	0.65	0.70	0.75	0.81	0.87	0.94	0.98	1.00
0.8	0.84	0.87	0.90	0.93	0.96	0.98	1.00	1.00
1	1.00	1.00	1.00	1.00	1.00	1.00	1.00	1.00
1.5	1.40	1.30	1.20	1.12	1.06	1.03	1.00	1.00
2	1.70	1.50	1.40	1.30	1.20	1.10	1.05	1.00
3	2.20	1.80	1.65	1.50	1.35	1.20	1.10	1.00
≥5	3.00	2.20	1.90	1.70	1.50	1.30	1.15	1.00

NOTES:
1. In-between values of S_c may be determined by straight-line interpolation from adjacent values.
2. For purposes of this table, D is the pipe OD.

Table 5-6 Values for the constrained modulus of the native soil at pipe elevation

Native in Situ Soils*						
		Cohesive				
Granular		q_u			M_{sn}	
Blows/ft† (0.3 m)	Description	*tons/ft²*	*kPa*	Description	*psi*	*MPa*
>0–1	Very, very loose	>0–0.125	0–13	Very, very soft	50	0.34
1–2	Very loose	0.125–0.25	13–25	Very soft	200	1.4
2–4		0.25–0.50	25–50	Soft	700	4.8
4–8	Loose	0.50–1.0	50–100	Medium	1,500	10.3
8–15	Slightly compact	1.0–2.0	100–200	Stiff	3,000	20.7
15–30	Compact	2.0–4.0	200–400	Very stiff	5,000	34.5
30–50	Dense	4.0–6.0	400–600	Hard	10,000	69.0
>50	Very dense	>6.0	>600	Very hard	20,000	138.0

* The constrained modulus, M_{sn}, for rock is ≥50,000 psi (≥345 MPa).
† Standard penetration test per ASTM D1586, Standard Test Method for Standard Penetration Test (SPT) and Split-Barrel Sampling of Soils.

NOTES:
1. For embankment installation $M_{sb} = M_{sn} = M_s$.
2. M_s special cases:
   - Solid sheeting—When permanent solid sheeting designed to last the life of the pipeline is used in the pipe zone, M_s shall be based solely on M_{sb}.
   - Cement-stabilized sand—When cement-stabilized sand is used as the embedment, initial deflections shall be based on a sand installation and the long-term M_{sb} = 25,000 psi (172 MPa). (Typical mix ratio is one sack of cement per ton or 1.5 sacks of cement per cubic yard of mix.)

5.7.4 Combined Loading

The maximum stress or strain resulting from the combined effects of internal pressure and deflection should meet Eqs 5-20 and 5-21 or Eqs 5-22 and 5-23 as follows:

For stress basis HDB:

$$\frac{\sigma_{pr} FS_{pr}}{HDB} + \frac{\sigma_b r_c}{S_b E} \le 1 \qquad \frac{\sigma_{pr} FS_{pr}}{HDB} + \frac{\sigma_b r_c}{S_b E \times 10^3} \le 1 \qquad \text{(Eq 5-20)}$$

$$\frac{\sigma_b r_c FS_b}{S_b E} + \frac{\sigma_{pr}}{HDB} \le 1 \qquad \frac{\sigma_b r_c FS_b}{S_b E \times 10^3} + \frac{\sigma_{pr}}{HDB} \le 1 \qquad \text{(Eq 5-21)}$$

For strain basis HDB:

$$\frac{\varepsilon_{pr} FS_{pr}}{HDB} + \frac{\varepsilon_b r_c}{S_b} \le 1 \qquad \frac{\varepsilon_{pr} FS_{pr}}{HDB} + \frac{\varepsilon_b r_c}{S_b} \le 1 \qquad \text{(Eq 5-22)}$$

$$\frac{\varepsilon_b r_c FS_b}{S_b} + \frac{\varepsilon_{pr}}{HBD} \le 1 \qquad \frac{\varepsilon_b r_c FS_b}{S_b} + \frac{\varepsilon_{pr}}{HBD} \le 1 \qquad \text{(Eq 5-23)}$$

Where:

FS_{pr} = pressure design factor, 1.8
FS_b = bending design factor, 1.5
σ_{pr} = hoop stress caused by internal working pressure, psi

$$= \frac{P_w D}{2t}$$

σ_b = bending stress caused by the maximum permitted deflection, psi

$$= D_f E \left(\frac{\delta d}{D}\right)\left(\frac{t_t}{D}\right)$$

r_c = rerounding coefficient, dimensionless
= $1 - P_w/435$ (where $P_w \le 435$ psi)
ε_{pr} = hoop strain caused by internal working pressure, in./in.

$$= \frac{P_w D}{2tE_H}$$

ε_b = bending strain due to maximum permitted deflection, in./in.

$$= D_f \left(\frac{\delta d}{D}\right)\left(\frac{t_t}{D}\right)$$

δd = maximum permitted long-term installed deflection, in.

Where:

FS_{pr} = pressure design factor, 1.8
FS_b = bending design factor, 1.5
σ_{pr} = hoop stress caused by internal working pressure, MPa

$$= \frac{P_w D}{2t}$$

σ_b = bending stress caused by the maximum permitted deflection, MPa

$$= 10^3 D_f E \left(\frac{\delta d}{D}\right)\left(\frac{t_t}{D}\right)$$

r_c = rerounding coefficient, dimensionless
= $1 - P_w/3{,}000$ (where $P_w \le 3{,}000$ kPa)
ε_{pr} = hoop strain caused by internal working pressure, mm/mm

$$= \frac{P_w D}{2tE_H}$$

ε_b = bending strain due to maximum permitted deflection, mm/mm

$$= D_f \left(\frac{\delta d}{D}\right)\left(\frac{t_t}{D}\right)$$

δd = maximum permitted long-term installed deflection, mm

5.7.5 Buckling

5.7.5.1 Buckling theory. Buried pipe is subjected to radial external loads composed of vertical loads and the hydrostatic pressure of groundwater and internal vacuum, if the latter two are present. External radial pressure sufficient to buckle buried pipe is many times higher than the pressure causing buckling of the same pipe in a fluid environment, because of the restraining influence of the soil.

5.7.5.2 Buckling calculations. The summation of appropriate external loads should be equal to or less than the allowable buckling pressure. The allowable buckling pressure q_a is determined by the following equation:

$$q_a = \frac{(1.2C_n)(EI)^{0.33}(\varphi_s M_s k_v)^{0.67} R_h}{(FS)r} \qquad q_a = \frac{(1.2C_n)(10^6 EI)^{0.33}(\varphi_s 10^3 M_s k_v)^{0.67} R_h}{(FS)r} \qquad \text{(Eq 5-24a)}$$

Where:

q_a = allowable buckling pressure, psi
FS = design factor, 2.5
C_n = scalar calibration factor to account for some nonlinear effects = 0.55
φ_s = factor to account for variability in stiffness of compacted embedment; suggested value is 0.9
$k\upsilon$ = modulus correction factor for Poisson's ratio, υ, of the soil
= $(1 + \upsilon)(1 - 2\upsilon)/(1 - \upsilon)$; in the absence of specific information, it is common to assume $\upsilon = 0.3$, giving $k\upsilon = 0.74$
R_h = correction factor for depth of fill
= $11.4/(11 + D/h)$
h = height of ground surface above top of pipe, in.

Where:

q_a = allowable buckling pressure, kPa
FS = design factor, 2.5
C_n = scalar calibration factor to account for some nonlinear effects = 0.55
φ_s = factor to account for variability in stiffness of compacted embedment; suggested value is 0.9
$k\upsilon$ = modulus correction factor for Poisson's ratio, υ, of the soil
= $(1 + \upsilon)(1 - 2\upsilon)/(1 - \upsilon)$; in the absence of specific information, it is common to assume $\upsilon = 0.3$, giving $k\upsilon = 0.74$
R_h = correction factor for depth of fill
= $11.4/(11 + D/1{,}000\,h)$
h = height of ground surface above top of pipe, m

An alternate form of Eq 5-24a is

$$q_a = (1/FS)\left[1.2C_n(0.149PS)^{0.33}\right](\varphi_s M_s k_v)^{0.67} R_h \qquad q_a = 1/FS\left[1.2C_n(0.149PS)^{0.33}\right](\varphi_s 10^3 M_s k_v)^{0.67} R_h \qquad \text{(Eq 5-24b)}$$

Typical pipe installations. Satisfaction of the buckling requirement is assured for typical pipe installations by using the following equation:

$$\gamma_w h_w + R_w(W_c) + P_\upsilon \le q_a \qquad [\gamma_w h_w + R_w(W_c)] \times 10^{-3} + P_\upsilon \le q_a \qquad \text{(Eq 5-25)}$$

Where:

γ_w = unit weight of water
= 0.0361 lb/in.^{3}
P_υ = internal vacuum pressure (i.e., the atmospheric pressure less absolute pressure inside pipe), psi
R_w = water buoyancy factor
= $1 - 0.33(h_w/h)$ $[0 \le h_w \le h]$
h_w = height of water surface above the pipe top, in.

Where:

γ_w = unit weight of water
= 9,800 N/m^3
P_υ = internal vacuum pressure (i.e., the atmospheric pressure less absolute pressure inside pipe), kPa
R_w = water buoyancy factor
= $1 - 0.33(h_w/h)$ $[0 \le h_w \le h]$
h_w = height of water surface above the pipe top, m

If live loads are considered, satisfaction of the buckling requirement is ensured by

$$\gamma_w h_w + R_w(W_c) + W_L \le q_a \qquad [\gamma_w h_w + R_w(W_c) + W_L] \times 10^{-3} \le q_a \qquad \text{(Eq 5-26)}$$

Typically, live load and internal vacuum are not considered simultaneously.

5.8 AXIAL LOADS

Three factors that contribute to the development of axial stresses in buried pipe are (1) hoop expansion as a result of internal pressure, which causes axial tensile stresses whenever the pipe is axially restrained; (2) restrained thermal expansion and contraction; and (3) pipe "beam" bending that may be induced by uneven bedding, differential soil settlement, or subsidence of soil. The minimum requirements for axial strengths are as specified by Sec. 5.1.2.4 and Sec. 5.1.2.5 and ANSI/AWWA C950, Tables 11, 12, and 13. These requirements include service conditions in typical underground pipe installations with unrestrained joints that comply with the guidelines provided in Chapter 6 of this manual and that have thrust blocks provided at bends, blanks, and valves in accordance with Chapter 7 and pipe manufacturers' recommendations. When restrained joints are used, the pipe should be designed to accommodate the full magnitude of forces generated by internal pressure.

5.9 SPECIAL DESIGN CONSIDERATIONS

Pipe that meets the design requirements of ANSI/AWWA C950 and Sec. 5.7 and that is installed in accordance with Chapter 6 guidelines has adequate strength for service in usual buried applications. Special consideration should be made for the following conditions: (1) elevated temperature service; (2) broad temperature fluctuations; (3) shallow burial, where H is less than 2 ft (0.6 m); (4) uneven bedding or differential settlement of unstable native soils; (5) restrained tension joints; (6) extremely difficult construction conditions (for example, subaqueous installation); and (7) unusually high surface or construction loads.

5.10 DESIGN EXAMPLE

Example design calculations are presented in this section for a specific situation. For reference, the set of design conditions, pipe properties, and installation parameters assumed for this design example are presented in Table 5-7. This summary is not repeated in the body of the example design calculations.

Table 5-7 Conditions and parameters for design example

Parameter	Value *in.-lb*	Units	Value *SI*	Units
	Design Conditions			
Nominal diameter	36	in.	900	mm
Working pressure	115	psi	800	kPa
Surge pressure	55	psi	375	kPa
Vacuum	14.7	psi	100	kPa
Cover depth				
Maximum	8	ft	2.4	m
Minimum	4	ft	1.2	m
Wheel load	20,000	lb	90,000	N
Service temperature	33–90	°F	1–32	°C
	Pipe Characteristics			
Pressure class	150	psi	1,000	kPa
Nominal stiffness	36	psi	248	kPa

(continued)

Table 5-7 Conditions and parameters for design example (*Continued*)

Parameter	Value *in.-lb*	Units	Value *SI*	Units
Inside diameter	36	in.	NA	NA
Outside diameter	NA	NA	924	mm
Total wall thickness	0.65	in.	16.5	mm
Liner thickness	0.04	in.	1	mm
Reinforced wall thickness	0.61	in.	15.5	mm
Mean diameter	36.69	in.	908.5	mm
Hoop tensile modulus	1,800,000	psi	12.5	GPa
HDB	0.0065	in./in.	0.0065	mm/mm
S_b	0.012	in./in.	0.012	mm/mm
Maximum permitted deflection	5	%	5	%
Installation Parameters				
Native soil				
Description	Dense silty sand			
Soil modulus	10,000	psi	69	MPa
Embedment				
Description	Moderately compacted sand (Class II at 90 percent compaction (PC))			
Soil modulus maximum depth	1,625	psi	11.2	MPa
Soil modulus minimum depth	1,500	psi	10.4	MPa
Unit weight	120	lb/ft^3	18,800	N/m^3
Groundwater effects				
Depth below grade	3	ft	0.9	m
Saturation reduction factor	0.85	NA	0.85	NA
Shape factor	5.5	NA	5.5	NA
Deflection lag factor	1.0	NA	1.0	NA
Deflection coefficient	0.1	NA	0.1	NA
Trench width	54	in.	1,350	mm

NA—not applicable

The pipe material properties and characteristics presented in Table 5-7 have been assumed for illustrative purposes and should not be used as actual design values. Values for these parameters differ for various pipe constructions and materials and should be obtained from the manufacturer.

Step 1. Confirm pressure class (Eq 5-2):

$$P_c \leq \frac{(HDB)(E_H)(2t)}{(FS)(D)} \text{ (strain basis)} \qquad P_c \leq \frac{(HDB)(E_H)(2t)10^6}{(FS)(D)} \text{ (strain basis)}$$

$$P_c = 150 \text{ psi} \leq \frac{(0.0065)(1.8 \times 10^6)(2)(0.61)}{1.8(36.69)} \qquad P_c = 1{,}000 \text{ kPa} \leq \frac{(0.0065)(12.5)(2)(15.5)(10^6)}{1.8(908.5)}$$

$$\leq 216 \text{ psi} \therefore \text{OK} \qquad \leq 1{,}537 \text{ kPa} \therefore \text{OK}$$

Step 2. Check working pressure (Eq 5-3):

$P_c \geq P_w$

150 psi ≥ 115 psi ∴ OK

$P_c \geq P_w$

1,000 kPa ≥ 800 kPa ∴ OK

Step 3. Check surge pressure (Eq 5-4):

$$P_c \geq \frac{(P_w + P_s)}{1.4}$$

$$P_c = 150 \text{ psi} \geq \frac{(115 + 55)}{1.4}$$

150 psi ≥ 122 psi ∴ OK

$$P_c \geq \frac{(P_w + P_s)}{1.4}$$

$$P_c = 1{,}000 \text{ kPa} \geq \frac{(800 + 375)}{1.4}$$

1,000 kPa ≥ 840 kPa ∴ OK

Step 4. Calculate limiting deflection (Eq 5-6):

ε_b = $(D_f)(\Delta y_a/D)(t_t/D) \leq (S_b)/FS$
$5.5(\Delta y_a/36.69)(0.65/36.69)$
$\leq (0.012)/1.5$
$0.00266\ \Delta y_a \leq 0.0080$
maximum Δy_a = 3.01 in.

limiting deflection
= (3.01/36.69) × 100
= 8.2%

permitted deflection = 5% ≤ 8.2% ∴ OK

ε_b = $(D_f)(\Delta y_a/D)(t_t/D) \leq (S_b)/FS$
$5.5(\Delta y_a/908.5)(16.5/908.5)$
$\leq (0.012)/1.5$
$0.00011\ \Delta y_a \leq 0.0080$
maximum Δy_a = 72.7 mm

limiting deflection
= (72.7/908.5) × 100
= 8.0%

permitted deflection = 5% ≤ 8.0% ∴ OK

Step 5. Calculate soils load (Eq 5-9): *(Maximum when the water table is below the pipe crown)*

$$W_c = \frac{\gamma_s H}{144}$$

At minimum depth of 4 ft,
W_c = 120(4)/144
= 3.33 psi

At maximum depth of 8 ft,
W_c = 120(8)/144
= 6.66 psi

$$W_c = \gamma_s H$$

At minimum depth of 1.2 m,
W_c = 18,800 (1.2)
= 22,560 N/m^2

At maximum depth of 2.4 m,
W_c = 18,800 (2.4)
= 45,120 N/m^2

Step 6. Calculate live loads (Eq 5-10):

$$W_L = \frac{M_p P I_f}{(L_1)(L_2)}$$

Determine impact factor (Eq 5-11)
I_f = 1 + 0.33[(96 – h)/96] ≥ 1.0

At minimum depth of 4 ft (48 in.)
I_f = 1 + 0.33[(96 – 48)/96] = 1.17

At maximum depth of 8 ft (96 in.)
I_f = 1 + 0.33[(96 – 96)/96] = 1.0

Determine L_1 (Eq 5-12)
L_1 = t_l + LLDF (h)

At minimum depth of 4 ft (48 in.)
L_1 = 10 + 1.15 (48) = 65 in.

At maximum depth of 8 ft (96 in.)
L_1 = 10 + 1.15 (96) = 120 in.

$$W_L = \frac{M_p P I_f}{(L_1)(L_2)}$$

I_f = 1 + 0.33[(2.4 – h)/2.4] ≥ 1.0

At minimum depth of 1.2 m
I_f = 1 + 0.33[(2.4 – 1.2)/2.4] = 1.17 ≥ 1.0

At maximum depth of 2.4 m
I_f = 1 + 0.33[(2.4 – 2.4)/2.4] ≥ 1.0

L_1 = t_l + LLDF (h)

At minimum depth of 1.2 m
L_1 = 0.254 + 1.15 (1.2) = 1.63 m

At maximum depth of 2.4 m
L_1 = 0.254 + 1.15 (2.4) = 3.01 m

Determine h_{int} (Eq 5-15)

h_{int} = (72 in. – t_w)/LLDF

h_{int} = (72 in. – 20)/1.15 = 45.2 in.

At minimum depth of 4 ft (48 in.)

$h > h_{int}$

then calculate L_2 (Eq 5-14)

L_2 = [t_w + 72 in. + LLDF (h)]/2

L_2 = [20 +72 + 1.15 (48)]/2 = 74 in.

At maximum depth of 8 ft (96 in.)

$h > h_{int}$

then calculate L_2 (Eq 5-14)

L_2 = [t_w + 72 in. + LLDF (h)]/2

L_2 = [20 + 72 + 1.15(96)]/2 = 101 in.

At minimum depth of 4 ft

$$W_L = \frac{1.2(20,000)(1.17)}{(65)(74)} = 5.84 \text{ psi}$$

At maximum depth of 8 ft

$$W_L = \frac{1.2(20,000)(1.0)}{(120)(101)} = 1.98 \text{ psi}$$

h_{int} = (1.83 m – t_w)/LLDF

h_{int} = (1.83 m – 0.508)/1.15 = 1.15 m

At minimum depth of 1.2 m

$h > h_{int}$

then calculate L_2 (Eq 5-14)

L_2= [t_w + 1.83 m + LLDF (h)]/2

L_2= [0.508 + 1.83 + 1.15(1.2)]/2 = 1.86 m

At maximum depth of 2.4 m

$h > h_{int}$

then calculate L_2 (Eq 5-14)

L_2= [t_w + 1.83 m + LLDF (h)]/2

L_2= [0.508 + 1.83 + 1.15(2.4)]/2 = 2.55 m

At minimum depth of 1.2 m

$$W_L = \frac{1.2(90,000)(1.17)}{(1.63)(1.86)} = 41,678 \text{ N/m}^2$$

At maximum depth of 2.4 m

$$W_L = \frac{1.2(90,000)(1.0)}{(3.01)(2.55)} = 14,071 \text{ N/m}^2$$

Step 7. Calculate the composite soil constrained modulus, M_s (Eq 5-19):

Determine M_{sb} (saturated):

At 4-ft cover,

Compute R_w = 1 – 0.33(1/4) = 0.918

Vertical soil pressure =

(120)(4)(0.918)/144 = 3.06 psi

From Table 5-4:

M_{sb} at 1 psi vertical stress = 1,275 psi

M_{sb} at 5 psi vertical stress = 1,500 psi

Interpolating for 3.06 psi vertical stress:

M_{sb} = 1,390 psi

Applying saturation reduction factor (0.85) yields:

M_{sb} = 1,181 psi

At 8-ft cover,

Compute R_w = 1 – 0.33(5/8) = 0.794

Vertical soil pressure =

(120)(8)(0.794)/144 = 5.29 psi

From Table 5-4:

M_{sb} at 5 psi vertical stress = 1,500 psi

M_{sb} at 10 psi vertical stress = 1,625 psi

Interpolating for 5.29 psi vertical stress:

M_{sb} = 1,507 psi

At 1.2-m cover,

Compute R_w = 1 – 0.33(0.3/1.2) = 0.918

Vertical soil pressure =

(18,800)(1.2)(0.918)/1000 = 22.6 kPa

From Table 5-4:

M_{sb} at 6.9 kPa vertical stress = 8.8 MPa

M_{sb} at 34.5 kPa vertical stress = 10.3 MPa

Interpolating for 22.6 kPa vertical stress:

M_{sb} = 9.65 MPa

Applying saturation reduction factor (0.85) yields:

M_{sb} = 8.20 MPa

At 2.4-m cover,

Compute R_w = 1 – 0.33(1.5/2.4) = 0.794

Vertical soil pressure =

(18,800)(2.4)(0.794)/1000 = 35.8 kPa

From Table 5-4:

M_{sb} at 34.5 kPa vertical stress = 10.3 MPa

M_{sb} at 69 kPa vertical stress = 11.2 MPa

Interpolating for 35.8 kPa vertical stress:

M_{sb} = 10.33 MPa

Applying saturation reduction factor (0.85) yields:

$M_{sb} = 1{,}281$ psi

M_{sn} = 10,000 psi
B_d = 54 in. at springline = 4.5 ft
D = 3.11 ft
B_d/D = 4.5/3.11 = 1.45

At 4 ft M_{sn}/M_{sb} = 10,000/1,181 = 8.47

At 8 ft M_{sn}/M_{sb} = 10,000/1,281 = 7.81

Interpolating from Table 5-5 yields:

S_c = 2.4
M_s = 2.4 (1,181) = 2,834 psi at 4-ft cover
M_s = 2.4 (1,281) = 3,074 psi at 8-ft cover

Applying saturation reduction factor (0.85) yields:

$M_{sb} = 8.8$ MPa

M_{sn} = 69 MPa
B_d = 1.35 m at springline
D = 0.924 m
B_d/D = 1.35/0.924 = 1.46

At 1.2 m M_{sn}/M_{sb} = 69/8.2 = 8.41

At 2.4 m M_{sn}/M_{sb} = 69/8.8 = 7.84

Interpolating from Table 5-5 yields:

S_c = 2.3
M_s = 2.3 (8.2) = 18.9 MPa at 1.2-m cover
M_s = 2.3 (8.8) = 20.2 MPa at 2.4-m cover

Step 8. Calculate the predicted deflection (Eq 5-8):

$$\frac{\Delta y}{D} = \frac{(D_L W_c + W_L)K_x}{0.149PS + 0.061M_s}$$

At 4-ft depth

$$= \frac{[1.0(3.33) + 5.84]0.1}{0.149(36) + 0.061(2{,}834)}$$

= 0.0051

= 0.51%

At 8-ft depth

$$= \frac{[1.0(6.66) + 1.98]0.1}{0.149(36) + 0.061(3{,}074)}$$

= 0.0045

= 0.45%

Predicted deflection is less than the maximum permitted deflection of 5% and the limiting deflection of 8.2% ∴ OK

$$\frac{\Delta y}{D} = \frac{(D_L W_c + W_L)K_x}{149PS + 61{,}000M_s}$$

At 1.2-m depth

$$= \frac{[1.0(22{,}560) + 41{,}678]0.1}{149(248) + 61{,}000(18.9)}$$

= 0.0054

= 0.54%

At 2.4-m depth

$$= \frac{[1.0(45{,}120) + 14{,}071]0.1}{149(248) + 61{,}000(20.2)}$$

= 0.0047

= 0.47%

Predicted deflection is less than the maximum permitted deflection of 5% and the limiting deflection of 8.0% ∴ OK

Step 9. Check combined loading (Eqs 5-22 and 5-23):

$$\varepsilon_{pr} = \frac{P_w D}{2t(E_{HT})}$$

$$\frac{115\,(36.69)}{2(0.61)(1.8 \times 10^6)}$$

= 0.0019 in./in.

$$\varepsilon_{pr} = \frac{P_w D}{2 \times 10^6\,(t)(E_{HT})}$$

$$\frac{800\,(908.5)}{2 \times 10^6\,(15.5)(12.5)}$$

= 0.0019 mm/mm

Calculate ε_b at the maximum permitted deflection using (Eq 5-6)

ε_b = $(D_f)(\delta d/D)(t_t/D)$
= 5.5(0.05)(0.65/36.69)
= 0.00487 in./in.
r_c = $1 - P_w/435$ = 1 – 115/435 = 0.73

ε_b = $(D_f)(\delta d/D)(t_t/D)$
= 5.5(0.05)(16.5/908.5)
= 0.0050 mm/mm
r_c = $1 - P_w/3{,}000$ = 1 – 800/3,000 = 0.73

$$\frac{\varepsilon_{pr}\,FS_{pr}}{HDB} + \frac{\varepsilon_b r_c}{S_b} \le 1$$

$$\frac{0.0019(1.8)}{0.0065} + \frac{(0.00487)(0.73)}{0.012} \le 1$$

$$0.82 \le 1 \ \therefore \text{OK}$$

$$\frac{\varepsilon_b r_c\,FS_b}{S_b} + \frac{\varepsilon_{pr}}{HDB} \le 1$$

$$\frac{0.00487(0.73)(1.5)}{0.012} + \frac{0.0019}{0.0065} \le 1$$

$$0.74 \le 1 \ \therefore \text{OK}$$

$$\frac{\varepsilon_{pr}\,FS_{pr}}{HDB} + \frac{\varepsilon_b r_c}{S_b} \le 1$$

$$\frac{0.0019(1.8)}{0.0065} + \frac{(0.0050)(0.73)}{0.012} \le 1$$

$$0.83 \le 1 \ \therefore \text{OK}$$

$$\frac{\varepsilon_b r_c\,FS_b}{S_b} + \frac{\varepsilon_{pr}}{HDB} \le 1$$

$$\frac{0.0050(0.73)(1.5)}{0.012} + \frac{0.0019}{0.0065} \le 1$$

$$0.75 \le 1 \ \therefore \text{OK}$$

Step 10. Check buckling (Eq 5-24a):

$$q_a = \frac{(1.2C_n)(EI)^{0.33}(\varphi_s M_s k_v)^{0.67} R_h}{(FS)r}$$

R_h = correction factor for depth of fill
= 11.4/ (11 + D/h)

At 4-ft (48-in.) depth
= 11.4/ (11 + 36.69/48)
= 0.97

At 8-ft (96-in.) depth
= 11.4/ (11 + 36.69/96)
= 1.00

Calculate EI using Eq 5-18

EI = PS (0.149)$[r + \Delta y_t/2]^3$
= 36 (0.149) $[36.69/2 + (0.05)(36.69)/2]^3$
= 38,336

At 4-ft depth

$$q_a = \frac{1.2(0.55)(38{,}336)^{0.33}\left[0.9(2{,}834)(0.74)\right]^{0.67}(0.97)}{2.5(36.69/2)}$$

= 71.9 psi

At 8-ft depth

$$q_a = \frac{1.2(0.55)(38{,}336)^{0.33}\left[0.9(3{,}074)(0.74)\right]^{0.67}(1.0)}{2.5(36.69/2)}$$

= 78.2 psi

$$q_a = \frac{(1.2C_n)(10^6 EI)^{0.33}(\varphi_s 10^3 M_s k_v)^{0.67} R_h}{(FS)r}$$

R_h = correction factor for depth of fill
= 11.4/ (11 + $D/1{,}000h$)

At 1.2-m depth
= 11.4/ (11 + 908.5/1,200)
= 0.97

At 2.4-m depth
= 11.4/ (11 + 908.5/2,400)
= 1.00

Calculate EI using Eq 5-18

EI = PS × 10^{-6}(0.149)$[r + \Delta y_t/2]^3$
= (248)(10^{-6})(0.149)$[908.5/2 + 0.05(908.5)/2]^3$
= 4,009

At 1.2-m depth

$$q_a = \frac{1.2(0.55)(10^6 \times 4{,}009)^{0.33}\left[0.9(18.9 \times 10^3)(0.74)\right]^{0.67}(0.97)}{2.5(908.5)/2}$$

= 485 kPa

At 2.4-m depth

$$q_a = \frac{1.2(0.55)(10^6 \times 4{,}009)^{0.33}\left[0.9(20.2 \times 10^3)(0.74)\right]^{0.67}(1.0)}{2.5(908.5)/2}$$

= 522 kPa

Check against requirement considering vacuum (Eq 5-25)

$\gamma_w h_w + R_w(W_c) + P_v \le q_a$

$[\gamma_w h_w + R_w(W_c)] \times 10^{-3} + P_v \le q_a$

R_w = water buoyancy factor	R_w = water buoyancy factor
= $1 - 0.33(h_w/h)$ $[0 \le h_w \le h]$	= $1 - 0.33(h_w/h)$ $[0 \le h_w \le h]$
At 4-ft (48-in.) depth	At 1.2-m depth
= 1 – 0.33 (12/48)	= 1 – 0.33 (0.3/1.2)
= 0.918	= 0.918
At 8-ft (96-in.) depth	At 2.4-m depth
= 1 – 0.33 (60/96)	= 1 – 0.33 (1.5/2.4)
= 0.794	= 0.794
Buckling check at 4-ft depth	Buckling check at 1.2-m depth
$0.0361(12) + 0.918(3.33) + 14.7 \le 71.9$	$[9{,}800(0.3) + 0.918(22{,}560)] \times 10^{-3} + 100 \le 485$
18.19 psi ≤ 71.9 psi ∴ OK	124 kPa ≤ 485 kPa ∴ OK
Buckling check at 8-ft depth	Buckling check at 2.4-m depth
$0.0361(60) + 0.794(6.66) + 14.7 \le 78.2$	$[9{,}800(1.5) + 0.794(45{,}120)] \times 10^{-3} + 100 \le 522$
22.15 psi ≤ 78.2 psi ∴ OK	151 kPa ≤ 522 kPa ∴ OK

Check against requirement considering live load (Eq 5-26)

$\gamma_w h_w + R_w(W_c) + W_L \le q_a$	$[\gamma_w h_w + R_w(W_c) + W_L] \times 10^{-3} \le q_a$
Buckling check at 4-ft depth	Buckling check at 1.2-m depth
$0.0361(12) + 0.918(3.33) + 5.84 \le 71.9$	$[9{,}800(0.3) + 0.918(22{,}560) + 41{,}678] \times 10^{-3} \le 485$
9.3 psi ≤ 71.9 psi ∴ OK	65.3 kPa ≤ 485 kPa ∴ OK
Buckling check at 8-ft depth	Buckling check at 2.4-m depth
$0.0361(60) + 0.794(6.66) + 1.98 \le 78.2$	$[9{,}800(1.5) + 0.794(45{,}120) + 14{,}071] \times 10^{-3} \le 522$
9.4 psi ≤ 78.2 psi ∴ OK	64.6 kPa ≤ 522 kPa ∴ OK

5.11 REFERENCES

AASHTO (American Association of State Highway and Transportation Officials). 2022. T 99. Standard Method of Test for Moisture-Density Relations of Soils Using a 2.5-kg (5.5-lb) Rammer and a 305-mm (12-in.) Drop. Washington, D.C.: AASHTO.

AASHTO. 2020 (9th ed.). *AASHTO LRFD Bridge Design Specifications*. Washington, D.C.: AASHTO.

ACPA (American Concrete Pipe Association). 2011. *Concrete Pipe Design Manual*. Irving, TX: ACPA.

ANSI/AWWA (American National Standards Institute/American Water Works Association). 2020. C950. Fiberglass Pressure Pipe. Denver: AWWA.

ASTM (ASTM International). 2022. D653. Standard Terminology Relating to Soil, Rock, and Contained Fluids. West Conshohocken, PA: ASTM.

ASTM. 2022. D1586/D1586M. Standard Test Method for Standard Penetration Test (SPT) and Split-Barrel Sampling of Soils. West Conshohocken, PA: ASTM.

ASTM. 2022. D3567. Standard Practice for Determining Dimensions of "Fiberglass" (Glass-Fiber-Reinforced Thermosetting Resin) Pipe and Fittings. West Conshohocken, PA: ASTM.

ASTM. 2021. D698. Standard Test Methods for Laboratory Compaction Characteristics of Soil Using Standard Effort (12,400 ft-lbf/ft^3 (600 kN-m/m^3)). West Conshohocken, PA: ASTM.

ASTM. 2021. D2412. Standard Test Method for Determination of External Loading Characteristics of Plastic Pipe by Parallel-Plate Loading. West Conshohocken, PA: ASTM.

ASTM. 2020. D2487. Standard Classification of Soils for Engineering Purposes (Unified Soil Classification System). West Conshohocken, PA: ASTM.

ASTM. 2020. D7382. Standard Test Methods for Determination of Maximum Dry Unit Weight of Granular Soils Using a Vibrating Hammer. West Conshohocken, PA: ASTM.

ASTM. 2019. D4253. Standard Test Methods for Maximum Index Density and Unit Weight of Soils Using a Vibratory Table. West Conshohocken, PA: ASTM.

Howard, A.K. 1996. *Pipeline Installation*. Lakewood, CO.: Relativity Publishing.

Howard, A.K. 1977. "Modulus of Soil Reaction Values for Buried Flexible Pipe." *Journal of Geotechnical Engineering*. 103(1).

McGrath, T.J. 1998. "Replacing É With the Constrained Modulus in Buried Pipe Design." In *Pipelines in the Constructed Environment* (J.P. Castronovo and J.A. Clark, editors). Reston, VA: American Society of Civil Engineers.

Portland Cement Association. 1950. *Vertical Pressure on Culverts Under Wheel Loads on Concrete Pavement Slabs*. Publication No. ST-65. Skokie, IL: Portland Cement Association.

Spangler, M.G. 1941. *The Structural Design of Flexible Pipe Culverts*. Bulletin 153. Ames, IA: Iowa State Engineering Experimental Station.

Chapter 6

Guidelines for Underground Installation of Fiberglass Pipe

6.1 INTRODUCTION

The structural and installation designs of fiberglass pipe, or most buried pipe, are closely related. The structural design process, discussed in Chapter 5, assumes that pipe will receive support from the embedment, and the installation process must ensure that the support is provided. The guidelines in this chapter suggest procedures for burial of fiberglass pipe in typically encountered soil conditions. Recommendations for trenching, placing, and joining pipe; placing and compacting embedment; and monitoring deflection levels are included.

ANSI/AWWA C950, Fiberglass Pressure Pipe, specifies pipe that encompasses a wide range of product variables. Diameters range from 1 to 156 in. (25–4,000 mm), and standard pipe stiffness classes range from 18 to 72 psi (124–496 kPa). There are nine pressure classes, which range from 50 to 450 psi (345–3,103 kPa) in 50-psi (345-kPa) increments. Engineers and installers should recognize that all possible combinations of pipe, embedment types, and natural ground conditions that may occur are not considered in this chapter. The recommendations provided may need to be modified or expanded to meet the needs of some installation conditions. Sec. 6.9 lists areas that may be influenced by project, local, or regional conditions and should be given consideration when preparing specifications. Guidance for installation of fiberglass pipe in subaqueous conditions is not included.

These guidelines are for use by engineers and specifiers, manufacturers, installation contractors, regulatory agencies, owners, and inspection organizations that are involved in the construction of buried fiberglass pipelines.

6.2 RELATED DOCUMENTS

The following ASTM standards provide engineers with additional information related to installing buried pipe.

D8	Standard Terminology Relating to Materials for Roads and Pavements
D420	Standard Guide to Site Characterization for Engineering, Design, and Construction Purposes (Withdrawn 2012)
D653	Standard Terminology Relating to Soil, Rock, and Contained Fluids
D698	Standard Test Methods for Laboratory Compaction Characteristics of Soil Using Standard Effort (12,400 ft-lbf/ft^3 (600 kN-m/m^3))
D883	Standard Terminology Relating to Plastics
D1556	Standard Test Method for Density and Unit Weight of Soil in Place by the Sand-Cone Method
D1586	Standard Test Method for Penetration Test (SPT) and Split-Barrel Sampling of Soils
D2167	Standard Test Method for Density and Unit Weight of Soil in Place by the Rubber Balloon Method (Withdrawn 2024)
D2216	Standard Test Method for Laboratory Determination of Water (Moisture) Content of Soil and Rock by Mass
D2321	Standard Practice for Underground Installation of Thermoplastic Pipe for Sewers and Other Gravity-Flow Applications
D2412	Standard Test Method for Determination of External Loading Characteristics of Plastic Pipe by Parallel-Plate Loading
D2487	Standard Practice for Classification of Soils for Engineering Purposes (Unified Soil Classification System)
D2488	Standard Practice for Description and Identification of Soils (Visual-Manual Procedure)
D3441	Standard Test Method for Mechanical Cone Penetration Tests of Soil
D3839	Standard Guide for Underground Installation of "Fiberglass" (Glass-Fiber-Reinforced Thermosetting-Resin) Pipe
D4253	Standard Test Methods for Maximum Index Density and Unit Weight of Soils Using a Vibratory Table
D4318	Standard Test Methods for Liquid Limit, Plastic Limit, and Plasticity Index of Soils
D4643	Standard Test Method for Determination of Water (Moisture) Content of Soil by the Microwave Oven Method
D4832	Standard Test Method for Preparation and Testing of Controlled Low Strength Material (CLSM) Test Cylinders
D4914	Standard Test Methods for Density and Unit Weight of Soil and Rock in Place by the Sand Replacement Method in a Test Pit
D4944	Standard Test Method for Field Determination of Water (Moisture) Content of Soil by the Calcium Carbide Gas Pressure Tester
D4959	Standard Test Method for Determination of Water (Moisture) Content of Soil by Direct Heating
D5030	Standard Test Method for Density of Soil and Rock in Place by the Water Replacement Method in a Test Pit
D5080	Standard Test Method for Rapid Determination of Percent Compaction

D5821	Standard Test Method for Determining the Percentage of Fractured Particles in Coarse Aggregate
D6103	Standard Test Method for Flow Consistency of Controlled Low Strength Material (CLSM)
D6024	Standard Test Method for Ball Drop on Controlled Low Strength Material (CLSM) to Determine Suitability for Load Application
D6938	Standard Test Methods for In-Place Density and Water Content of Soil and Soil-Aggregate by Nuclear Methods (Shallow Depth)
D7382	Standard Test Methods for Determination of Maximum Dry Unit Weight of Granular Soils Using a Vibrating Hammer
F412	Standard Terminology Relating to Plastic Piping Systems
F1668	Standard Guide for Construction Procedures for Buried Plastic Pipe

6.3 TERMINOLOGY

Terminology used in this chapter is in accordance with ASTM Standards D8, D653, D883, and F412 unless otherwise indicated. The following terms are specific to this manual.

Backfill. Backfill material placed from the top of the embedment to the ground surface (see Figure 6-1).

Bedding. Backfill material placed in the bottom of the trench or on the foundation to provide a uniform material on which to lay the pipe; the bedding may or may not include part of the haunch zone (see Figure 6-1).

Compacted. Densification of soil using mechanical compaction equipment or using water.

Deflection. Any change in the diameter of the pipe resulting from installation and imposed loads. Deflection may be measured and reported as change in either vertical or horizontal diameter and is usually expressed as a percentage of the undeflected pipe diameter.

Embedment. Backfill material placed at the sides of the pipe and up to 6 to 12 in. (150–300 mm) over the top of the pipe, including the haunching (see Figure 6-1).

Engineer. The engineer or the duly recognized or authorized representative in responsible charge of the work.

Fines. Soil particles that pass a No. 200 (0.076-mm) sieve.

Foundation. Material placed and compacted in the bottom of the trench to replace overexcavated material and/or to stabilize the trench bottom if unsuitable ground conditions are encountered (see Figure 6-1).

Geotextile. Any permeable textile material used with foundation, soil, earth, rock, or any other geotechnical engineering-related material as an integral part of a synthetic product, structure, or system.

Haunch zone. Embedment material placed on top of the bedding and under the springline of the pipe; the term only pertains to soil directly beneath the pipe (see Figure 6-1).

Manufactured aggregates. Aggregates such as slag that are products or byproducts of a manufacturing process or natural aggregates that are reduced to their final form by a manufacturing process such as crushing.

Native (in situ) soil. Natural soil in which a trench is excavated for pipe installation or on which a pipe and embankment are placed.

Open-graded aggregate. An aggregate that has a particle size distribution such that, when compacted, the resulting voids between the aggregate particles are relatively large.

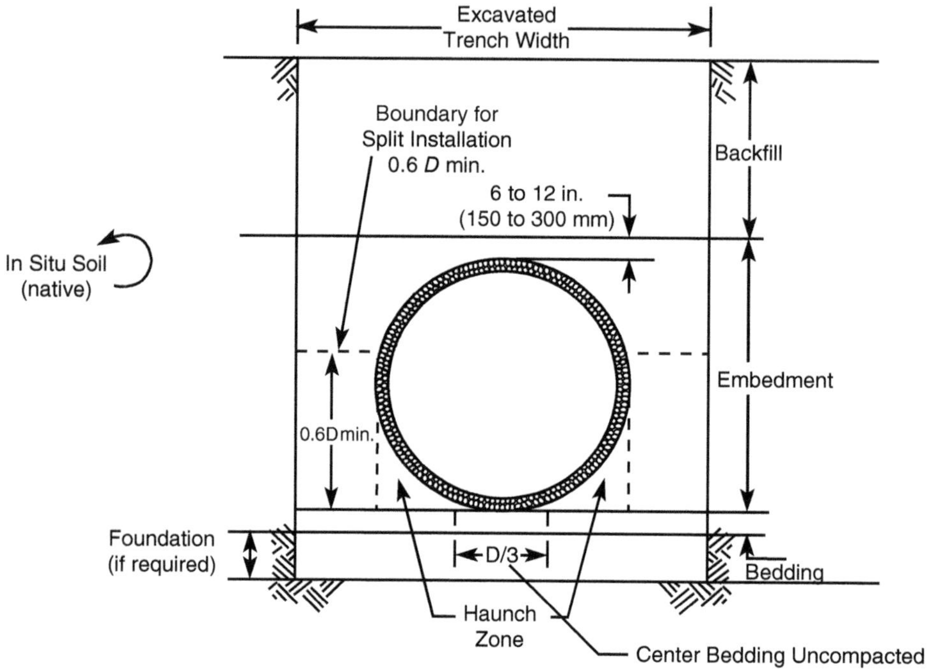

Figure 6-1 Trench cross-section terminology

Optimum moisture content. The moisture content of soil at which its maximum density is obtained (see ASTM D698).

Percent compaction (PC). See Sec. 5.2.1.

Processed aggregates. Aggregates that are screened, washed, mixed, or blended to produce a specific particle size distribution.

Soil stiffness. A property of soil, generally represented numerically by a modulus of deformation, that indicates the relative amount of deformation that will occur under a given load.

Split installation. An installation where the embedment is composed of two different materials or one material placed at two different densities. The lower material extends from the top of the bedding to a depth of at least 0.6 times the diameter.

Uncompacted. Placement of soil that is not compacted but is worked with hand tools, such as shovels or rods, to ensure all voids are filled.

6.4 IN SITU SOILS

It is important to understand in situ conditions before construction to prepare proper specifications and plan construction methods. Classification of soils according to ASTM D2487 and ASTM D2488 is useful in understanding in situ conditions. Other tests, like the standard penetration and cone penetrometer tests, are also useful in determining soil stiffness. Depending on actual installation conditions, such as trench geometry, the in situ soil conditions may also have a significant effect on pipe design. Refer to Chapter 5 for further discussion.

Consideration should also be given to seasonal or long-term variations in groundwater level when evaluating groundwater conditions. For example, if the soil exploration program is conducted in August, the groundwater level may be quite low compared with levels in April or May.

6.5 EMBEDMENT MATERIALS

Soil types used or encountered when burying pipe include those classified in Chapter 5, Table 5-3, as well as natural, manufactured, and processed aggregates. Soils are grouped into categories in Table 6-1 based on their typical soil stiffness when compacted. Class I indicates a soil that provides the highest stiffness with the least compactive energy. Each higher number soil class is successively less compactible, i.e., it provides less soil stiffness at a given PC and requires greater compactive energy to provide a given level of stiffness. See Chapter 5 for a discussion of how soil stiffness affects buried pipe behavior.

Table 6-2 provides recommendations on installation and use of embedment materials based on soil class and location in the trench. In general, soil conforming to Class I through Class IV should be used as recommended, and Class V materials should be excluded from the embedment.

6.5.1 Soil Stiffness Classes

Class I. Class I materials provide maximum pipe support due to a low content of sand and fines. With minimum effort, these materials can be installed at relatively high soil stiffness over a wide range of moisture contents. Even if uncompacted, Class I materials provide good support if they are placed and worked into the haunch zone to fill all voids. In addition, the high permeability of Class I materials may aid in the control of water, making them desirable for embedment in rock cuts where water is frequently

Table 6-1 Soil classes

Soil Classes	Unified Soil Classification System Soil Groups*	American Association of State Highway and Transportation Officials (AASHTO) Soil Groups†
Class I	Crushed rock: ≤15% sand, maximum 25% passing the ⅜-in. sieve and maximum 5% passing No. 200 sieve‡	
Class II	Clean, coarse-grained soils: SW, SP, GW, GP, or any soil beginning with one of these symbols with 12% or less passing No. 200 sieve§	A1, A3
Class III	Coarse-grained soils with fines: GM, GC, SM, SC, or any soil beginning with one of these symbols with more than 12% fines	A-2-4, A-2-5, A-2-6, or A-4 or A-6 soils with more than 30% retained on a No. 200 sieve
	Sandy or gravelly fine-grained soils: CL, ML (or CL-ML, CL/ML, ML/CL) with more than 30% retained on a No. 200 sieve	
Class IV	Fine-grained soils: CL, ML (or CL-ML, CL/ML, ML/CL) with 30% or less retained on a No. 200 sieve	A-2-7, or A-4 or A-6 soils with 30% or less retained on a No. 200 sieve
Class V	Highly plastic and organic soils: MH, CH, OL, OH, PT	A5, A7

NOTES:
* ASTM D2487, Standard Classification of Soils for Engineering Purposes (Unified Soil Classification System).
† AASHTO M145, Standard Specification for Classification of Soils and Soil-Aggregate Mixtures for Highway Construction Purposes.
‡ Crushed rock is defined as angular and subangular per ASTM D2488 and D5821. Class I crushed rock particles shall have all fractured faces.
§ Uniform fine sands (SP) with more than 50% passing a No. 100 sieve (0.006 in. [0.15 mm]) are very sensitive to moisture and should not be used as backfill for fiberglass pipe unless specifically allowed in the contract documents. If use of these materials is allowed, compaction and handling procedures should follow the guidelines for Class III materials.

Table 6-2 Recommendations for installation and use of soils and aggregates for foundation and embedment

Soil Classes*	Class I	Class II	Class III	Class IV
General recommendations and restrictions	Acceptable and common where no migration is probable or when combined with a geotextile filter media. Suitable for use as a drainage blanket and underdrain where adjacent material is suitably graded or when used with a geotextile filter fabric (see Sec. 6.5.2).	Where hydraulic gradient exists, check gradation to minimize migration. Clean groups are suitable for use as a drainage blanket and underdrain (see Table 5-3). Uniform fine sands (SP) with more than 50% passing a No. 100 sieve (0.006 in. [0.15 mm]) behave like silts and should be treated as Class III soils.	Do not use where water conditions in trench prevent proper placement and compaction. Not recommended for use with pipes with stiffness of 9 psi (62 kPa) or less.	Difficult to achieve high soil stiffness (see Sec. 6.5.1). Do not use where water conditions in trench prevent proper placement and compaction. Not recommended for use with pipes with stiffness of 9 psi (62 kPa) or less.
Bedding	Suitable as bedding and for replacing overexcavated and unstable trench bottom as restricted above.	Suitable as bedding and for replacing overexcavated and unstable trench bottom as restricted above.	Suitable for replacing overexcavated trench bottom as restricted above.	Not suitable.
Embedment	Suitable as restricted above. Work material under pipe to provide uniform haunch support.	Suitable as restricted above. Work material under pipe to provide uniform haunch support. Install and compact in 12-in. (300-mm) maximum layers.	Suitable as restricted above. Difficult to place and compact in the haunch zone. Install and compact in 6-in. (150-mm) maximum layers.	Suitable as restricted above. Difficult to place and compact in the haunch zone.
Embedment compaction				
Minimum recommended percent compaction	Uncompacted, i.e., careful placement to fill all voids	85% (ASTM D4253)	90% (ASTM D698)	95% (ASTM D698)
Relative compactive effort required to achieve required compaction	Low	Moderate	High	Very high
Compaction methods	Vibration or impact	Vibration or impact	Impact	Impact
Required moisture control	None	None	Maintain near optimum to minimize compactive effort	Maintain near optimum to minimize compactive effort

* Class V materials are unsuitable as embedment. They may be used as backfill as permitted by the engineer.

encountered. However, when groundwater flow is anticipated, consideration should be given to the potential for migration of fines from adjacent materials into the open-graded Class I material (see Sec. 6.5.2).

Class II. Compacted Class II materials provide a relatively high level of pipe support. However, open-graded groups may allow migration, and the sizes should be checked for compatibility with adjacent material (see Sec. 6.5.2).

Class III. Class III materials provide less support for a given PC than Class I or Class II materials. Higher levels of compactive effort are required, and moisture content must be near optimum to minimize compactive effort and achieve the required density. These materials provide reasonable levels of pipe support once proper density is achieved.

Class IV. Class IV materials require a geotechnical evaluation before use. The moisture content must be near optimum to minimize compactive effort and achieve the required density. When properly placed and compacted, Class IV materials can provide reasonable levels of pipe support; however, these materials may not be suitable under high fills, surface-applied wheel loads, or high energy–level vibratory compactors and tampers. Do not use where water conditions in the trench prevent proper placement and compaction.

Class V. Class V materials are not suitable for use as embedment for flexible pipe.

6.5.2 Considerations for Use of Soil in Backfill

Moisture content of embedment materials. The moisture content of embedment materials with substantial fines must be controlled to permit placement and compaction to required levels. For soils with low permeability (i.e., Class III, Class IV, and some dual-symbol Class II soils), moisture content is typically controlled to ±3 percent of optimum (ASTM D698). Obtaining and maintaining the required limits on moisture content are important criteria for selecting materials because failure to achieve required density, especially for the embedment, may result in excessive deflection.

Class I and most Class II materials are free-draining and require little or no control of moisture for compaction.

Compatibility of pipe and embedment. Experience has shown that pipe deflections and strain levels increase when low-stiffness pipe is embedded in materials that require large compactive efforts. This occurs because of the local distortions of the pipe shape that result as compactive energy is applied to the embedment. Because of this, it is recommended that pipe with stiffness of 9 psi (62 kPa) or less be embedded only in soil type Class I or Class II.

Maximum particle size. Maximum particle size for embedment should be 1 to 1.5 times the pipe wall thickness with a maximum of 1.5 in. For backfill, the maximum particle size allowed should not exceed 75 percent of the lift thickness. When backfill contains cobbles, boulders, etc., the embedment should be extended above the top of the pipe at least 12 in. (300 mm). Backfill containing boulders larger than 8 in. (200 mm) shall not be dropped or rolled down sloping trench walls onto the backfill from a height greater than 6 ft (1.8 m) until the depth of fill over the top of the pipe is greater than 24 in. (600 mm).

Migration. When open-graded material is placed adjacent to finer material, fines may migrate into the coarser material under the action of hydraulic gradient from groundwater flow. Significant hydraulic gradients may arise in the pipeline trench during construction when water levels are controlled by various pumping or well-pointing methods or after construction when permeable underdrain or embedment materials act as a French drain under high groundwater levels. Field experience shows that migration can result in significant loss of pipe support and increasing deflections that may eventually exceed design limits. The gradation and relative size of the embedment and adjacent materials must be compatible to minimize migration. In general, where significant groundwater flow is anticipated, avoid placing coarse open-graded materials such as Class I above, below, or adjacent to finer materials unless methods are used to impede migration. For example, consider the use of an appropriate soil filter or a geotextile filter fabric along the boundary of the incompatible materials.

The following filter gradation criteria may be used to restrict migration of fines into the voids of coarser material under a hydraulic gradient:

- $D_{15}/d_{85} < 5$, where D_{15} is the sieve opening size passing 15 percent by weight of the coarser material, and d_{85} is the sieve opening size passing 85 percent by weight of the finer material.
- $D_{50}/d_{50} < 25$, where D_{50} is the sieve opening size passing 50 percent by weight of the coarser material, and d_{50} is the sieve opening size passing 50 percent by weight of the finer material. (This criterion need not apply if the coarser material is well-graded [see ASTM D2487]).

If the finer material is a medium to highly plastic clay (CL or CH), the following criterion may be used in lieu of the D_{15}/d_{85} criteria: $D_{15} < 0.02$ in. (0.5 mm), where D_{15} is the sieve opening size passing 15 percent by weight of the coarser material.

These criteria may need to be modified if one of the materials is gap graded. Materials selected for use based on filter gradation criteria should be handled and placed in a manner that will minimize segregation.

Cementitious embedment materials. Embedment materials supplemented with cement to improve long-term strength and/or stiffness (soil cement, cement-stabilized backfill) or to improve flowability (flowable fill, controlled low-strength material) have been shown to be effective embedment materials in terms of ease of placement and quality of support to pipe.

Flowable Fill. Flowable fill (ASTM refers to this material as CLSM) is a fluid mixture of Portland cement, soil, and water that hardens into a solid mass. Flowable fill is typically about two to five times stiffer than compacted soil and thus provides good support for buried pipe. Flowable fill can range from material from a concrete batch plant to a mixture using the native soils excavated from the trench. The unconfined compressive strength of flowable fill should be between 50 and 100 psi at 28 days (ASTM D4832). At this strength, the material is easily excavated for repairs or future work. To ensure that the material will flow and completely fill the space between the pipe and the trench soil, the mixture should have a spread of 8 to 12 in. (ASTM D6103). The strength of the material depends on the amount of cementitious material, the aggregate, and the time after placement. Unless it is a high early strength mixture, flowable fill should not be backfilled until the next day after placement. ASTM D6024 describes a method to evaluate when backfilling or paving can be started.

Flowable fill can be used for the embedment for either basic or engineered installation. The height of flowable fill can be to the quarter point of the outside of the pipe to provide good haunch support to the three-quarter point to provide structural support for the pipe or to 12 in. above the pipe for full embedment. If the flowable fill is used just for haunch support, the remainder of the embedment up to the three-quarter point should be placed at 95 percent or greater (ASTM D698). Flowable fill can also be used as the backfill under roads or other paved areas. However, flowable fill should never be used as backfill unless it is also used as the embedment material.

The pipe may float because flowable fill is placed as a fluid. Restraints will be necessary, or the material should be placed in lifts. The height of the first lift should be just below the level where the pipe will float. After the initial set, typically three to five hours, the remaining flowable fill can be placed. The adhesion between the flowable fill and the pipe may prevent flotation. This will not work if the bond is broken, as in the case of high temperature differentials.

Flowable fill is a sustainable material in which the ingredients can be waste materials or byproducts. For example, Class C fly ash or cement kiln dust can be used instead of Portland cement, and some projects use the soil excavated from the trench as the aggregate in the mixture. Nonpotable water can be used.

6.6 TRENCH EXCAVATION

6.6.1 Excavation

Excavate trenches to ensure that sides will be stable under all working conditions. Slope trench walls or provide support in conformance with safety standards. Open only enough trench that can be safely maintained by available equipment. Place and compact embedment and backfill in trenches as soon as practicable, preferably no later than the end of each working day. Place excavated material away from the edge of the trench to minimize the risk of trench wall collapse.

Water control. It is always good practice to remove water from a trench before laying and embedding pipe. Although circumstances occasionally require pipe installation in conditions of standing or running water, such practice is outside the scope of this chapter. Prevent runoff and surface water from entering the trench at all times.

Groundwater. When groundwater is present in the work area, dewater to maintain stability of in situ and imported materials. Maintain water level below pipe bedding. Use sump pumps, wellpoints, deep wells, geotextiles, perforated underdrains, or stone blankets of sufficient thickness to remove and control water in the trench. When excavating, ensure that the groundwater is below the bottom of the cut at all times to prevent washout from behind sheeting or sloughing of exposed trench walls. Maintain control of water in the trench before, during, and after pipe installation and until embedment is installed and sufficient backfill has been placed to prevent flotation of the pipe (see Sec. 6.7.3). To preclude loss of soil support, use dewatering methods that minimize removal of fines and the creation of voids within in situ materials.

Running water. Control running water that emanates from surface drainage or groundwater to preclude undermining of the trench bottom or walls, the foundation, and bedding or embedment. Provide dams, cutoffs, or other barriers at regular intervals along the installation to preclude transport of water along the trench bottom. Backfill all trenches as soon as practical after the pipe is installed to prevent disturbance of pipe and embedment.

Materials for water control. Use suitably graded materials for foundation layers to transport running water to sump pits or other drains. Use properly graded materials and/or perforated underdrains to enhance transport of running water as required. Select the gradation of the drainage materials to minimize migration of fines from surrounding materials (see Sec. 6.5.2).

Minimum trench width. Where trench walls are stable or supported, provide a width sufficient but not greater than necessary to ensure working room to properly and safely place and compact haunching and embedment materials. The space between the pipe and trench wall must be 6 in. (150 mm) wider than the compaction equipment used in this region. For a single pipe in a trench, minimum width at the bottom of the trench should be 1.25 times the outside diameter of the pipe plus 12 in. (300 mm). For multiple pipes in the same trench, clear space between pipes must be at least the average of the radii of the two adjacent pipes for depths greater than 12 ft (3.5 m) and two-thirds of the average of the radii of the two adjacent pipes for depths less than 12 ft (3.5 m). The distance from the outside pipe to the trench wall must not be less than if that pipe were installed as a single pipe in a trench. If mechanical compaction equipment is used, the minimum space between pipe and trench wall or between adjacent pipe shall not be less than the width of the widest piece of equipment plus 6 in. (150 mm).

In addition to safety considerations, the trench width in unsupported, unstable soils will depend on the size and stiffness of the pipe, stiffness of the embedment and in situ soil, and depth of cover. Specially designed equipment or the use of free-flowing backfill, such as uniform rounded pea gravel or flowable fill, may enable the satisfactory installation and embedment of pipe in trenches narrower than previously specified. If the use of such equipment or backfill material provides an installation consistent with the requirements of this manual, minimum trench widths may be reduced if approved by the engineer.

Support of trench walls. When supports such as trench sheeting, trench jacks, or trench shields or boxes are used, ensure that support of the embedment is maintained throughout the installation process. Ensure that sheeting is sufficiently tight to prevent washing out of the trench wall from behind the sheeting. Provide tight support of trench walls below viaducts, existing utilities, or other obstructions that restrict driving of sheeting.

Supports left in place. Sheeting driven into or below the top of the embedment should be left in place to preclude loss of support of foundation and embedment materials. When the top of the sheeting is to be cut off, make the cut 1.5 ft (0.5 m) or more above the crown of the pipe. Leave walers and braces in place as required to support cutoff sheeting and the trench wall in the vicinity of the pipe. Timber sheeting to be left in place is considered a permanent structural member and should be treated against biological degradation (e.g., attack by insects or other biological forms) as necessary and against decay if above groundwater. Note that certain preservative and protective compounds may pose environmental hazards. Determination of acceptable compounds is outside the scope of this manual.

Movable trench wall supports. Do not disturb the installed pipe or the embedment when using movable trench boxes and shields. Movable supports should not be used below the top of the embedment unless approved methods are used for maintaining the integrity of embedment material. Before moving supports, place and compact embedment to sufficient depths to ensure protection of the pipe. As supports are moved, finish placing and compacting embedment.

Removal of trench wall support. If the removal of sheeting or other trench wall supports that extend below the top of the pipe is permitted, ensure that pipe, foundation, and embedment materials are not disturbed by support removal. Fill voids left after removal of supports and compact all material to required densities. Pulling the trench wall support in stages as backfilling progresses is advised.

6.6.2 Trench Bottom

Excavate trenches to specified grades. See Sec. 6.7.1 for guidance on installing foundation and bedding.

Excavate the trench a minimum of 4 in. (100 mm) below the bottom of the pipe. When ledge, rock, hardpan, or other unyielding material or cobbles, rubble, debris, boulders, or stones larger than 1.5 in. (40 mm) are encountered in the trench bottom, excavate a minimum depth of 6 in. (150 mm) below the pipe bottom.

If the trench bottom is unstable or shows a "quick" tendency, overexcavate as required to provide the proper foundation.

The native material may be used for bedding and embedment if it meets all criteria of the specified embedment materials. Trench preparation is discussed in Sec. 6.7.1.

6.6.3 Trenching on Slopes

The angle at which slopes can become unstable depends on the quality of the soil. The risk of unstable conditions increases dramatically with slope angle. In general, pipes should not be installed on slopes greater than 15° (a slope of 1:4) or in areas where slope instability is suspected unless supporting conditions have been verified by a proper geotechnical investigation. Installing pipes aboveground may be a preferred method for steep slopes because aboveground structures, such as pipe supports, are more easily defined and therefore, the quality of installation is easier to monitor and settlement easier to detect. Pipes may be installed on slopes greater than 15° (a slope of 1:4), provided that the following conditions are met.

- Long-term stability of the installation can be ensured with proper geotechnical design.
- Pipes are embedded with coarse-grained Class I material with high shear strength, or the shear strength of the backfill is ensured by other means. Compact the embedment in accordance with the requirements of Sec. 6.7.3.
- Pipes are installed in straight alignment (±0.2°) with minimum gap between pipe ends.
- Absolute long-term movement of the embedment in the axial direction of the pipe is less than 0.75 in. (20 mm) to avoid joint separation.
- The installation is properly drained to avoid washout of materials and ensure adequate soil shear strength. This may include treatment in the backfill or on the ground surface.
- Stability of individual pipes is monitored throughout the construction phase and the first stages of operation.
- The manufacturer is consulted to determine whether a special pipe design is required.

6.7 PIPE INSTALLATION

Recommendations for use of the various types of materials classified in Sec. 6.5.1 and Chapter 5, Table 5-2 for foundation, bedding, haunching, and embedment are provided in Table 6-2. Installation of pipe in areas where significant settlement may be anticipated—such as in backfill adjacent to building foundations, sanitary landfills, or in other highly unstable soils—requires special engineering and is outside the scope of this manual.

6.7.1 Preparation of Trench

Foundation and bedding. Provide a firm, stable, and uniform support for the pipe barrel and any protruding features of its joint (see Figure 6-2). Provide a minimum of 4 in. (100 mm) of bedding below the barrel and 3 in. (75 mm) below any part of the pipe like expanded bells, unless otherwise specified.

Bedding material. In general, the bedding material will need to be an imported material to provide the proper gradation and pipe support. It is preferable that the same material be used for the embedment. To determine whether the native material is acceptable as bedding material, it should meet all requirements of the embedment. This determination must be made constantly during the pipe installation process because native soil conditions vary widely and change suddenly along the length of a pipeline. It is becoming common practice to leave the bedding uncompacted for a width of one-third of the pipe diameter centered directly under the pipe. This reduces concentrated loads on the invert (Figure 6-1).

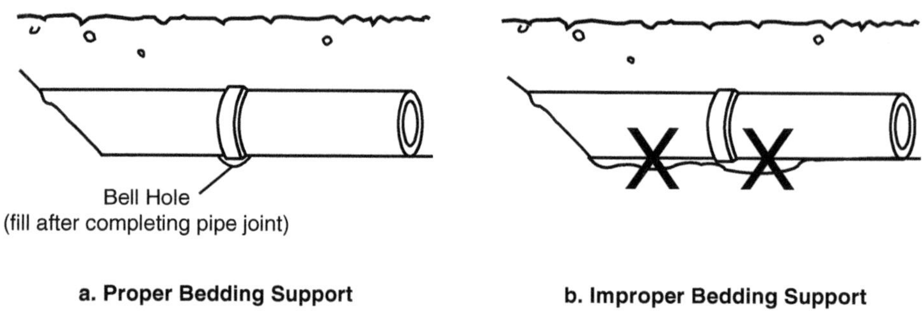

Source: Flowtite Technology, Sandefjord, Norway

Figure 6-2 Examples of bedding support

Rock and unyielding materials. When rock or unyielding material is present in the trench bottom, install a cushion of bedding of 6-in. (150-mm) minimum thickness below the bottom of the pipe. If a sudden transition from rock to a softer material under the pipe occurs, steps must be taken to accommodate possible differential settlement. Figure 6-3(b) illustrates one method; however, other methods are also possible.

Unstable trench bottom. Where the trench bottom is overexcavated because of unstable or quick conditions, install a foundation of Class I or Class II material. Use a suitably graded material where conditions may cause migration of fines and loss of pipe support. Place and compact foundation material in accordance with Table 6-2. For severe conditions, a special foundation, such as piles or sheeting capped with a concrete mat, may be required. The use of appropriate geotextiles can control "quick" and unstable trench bottom conditions.

Localized loadings. Minimize localized loadings and differential settlement wherever the pipe crosses other utilities or subsurface structures (see Figures 6-3 and 6-4) or whenever special foundations are used, such as concrete-capped piles or sheeting. Provide a 12-in. (300-mm) minimum cushion of bedding or embedment between the pipe and any point of localized loading.

Overexcavation. If the trench bottom is excavated below intended grade, fill the overexcavation with compatible foundation or bedding material and compact to a density not less than the minimum densities listed in Table 6-2.

Sloughing. If trench sidewalls slough off during any excavation or installation of embedment, remove all sloughed and loose material from the trench.

6.7.2 Placing and Joining Pipe

Location and alignment. Place pipe and fittings in the trench with the invert conforming to the required elevations, slopes, and alignment. Provide bell holes in pipe bedding no larger than necessary to ensure uniform pipe support. Fill all voids under the bell by working in bedding material. In special cases in which the pipe is to be installed to a curved alignment, maintain angular joint deflection (axial alignment) and pipe bending radius within acceptable design limits. Pipe should be laid on flat, uniform material that is at the appropriate grade. Do not bring pipe to grade by using mounds of soil or other material at discreet points along the length of the pipe. When pipe laying is interrupted, secure piping against movement and seal open ends to prevent the entrance of water, mud, or foreign material.

Jointing. Comply with manufacturer's recommendations for assembly of joint components, lubrication, and making of joints.

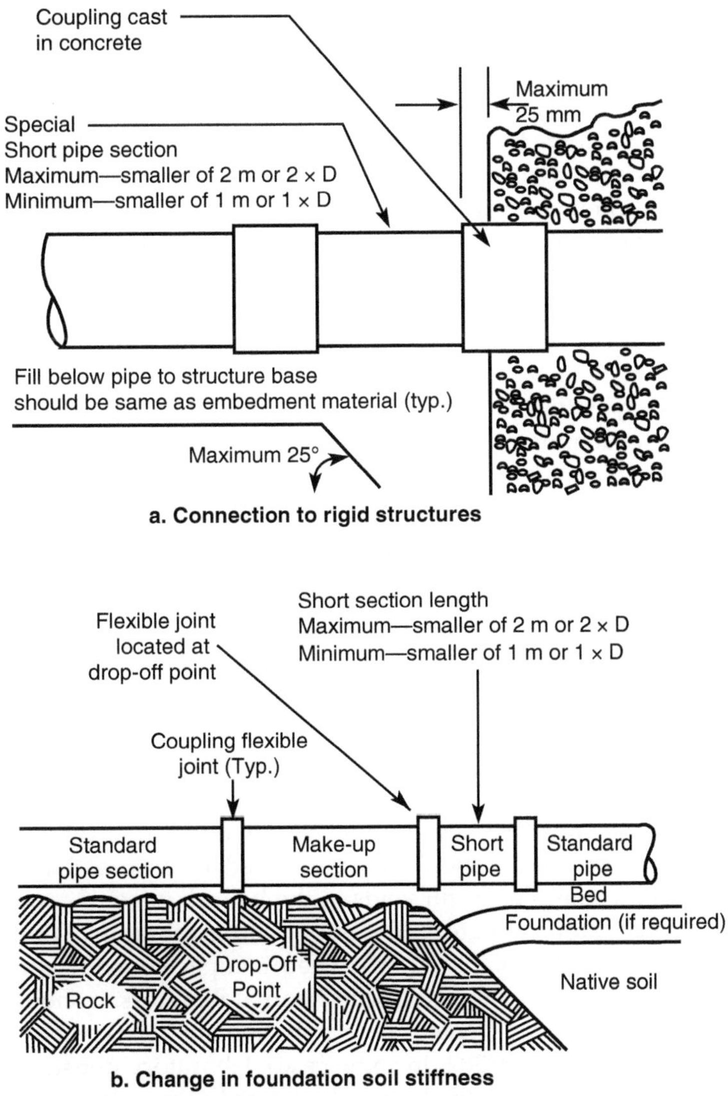

Source: Adapted from Flowtite Technology, Sandefjord, Norway

Figure 6-3 Accommodating differential settlement

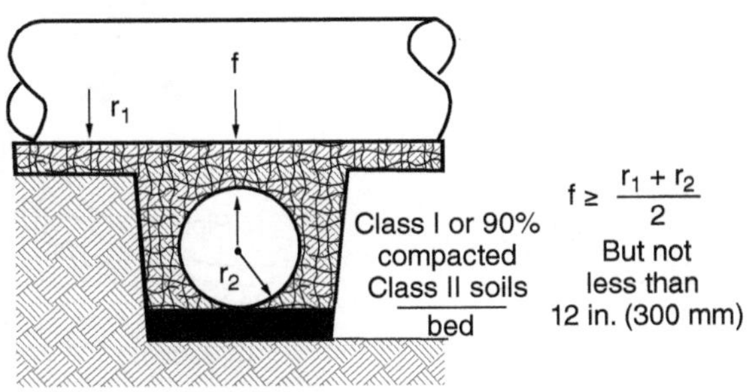

Source: Adapted from Flowtite Technology, Sandefjord, Norway

Figure 6-4 Crossover of adjacent piping systems

Elastomeric seal (gasketed) joints. Mark pipe ends or verify that pipe ends are marked to indicate insertion stop position and that pipe is inserted into pipe or fitting bells to this mark. Push spigot into bell using methods recommended by the manufacturer, keeping pipe true to line and grade. Protect the end of the pipe during homing, and do not use excessive force that may result in overassembled joints or dislodged gaskets. If full entry is not achieved, disassemble and clean the joint and reassemble. Use only lubricant supplied or recommended for use by the pipe manufacturer. Do not exceed manufacturer's recommendations for angular deflection (axial alignment).

Adhesive-bonded and wrapped joints. When making adhesive-bonded and wrapped joints, follow recommendations of the pipe manufacturer. Allow freshly made joints to set for the recommended time before moving, burying, or otherwise disturbing the pipe.

Angularly deflected joints. Large radius bends in pipelines may be accomplished by rotating the alignment of adjacent lengths of pipe (i.e., angularly deflecting the joint). The amount of angular deflection should not exceed the manufacturer's recommendations.

6.7.3 Placing and Compacting Pipe Backfill Materials

Place embedment materials by methods that will not disturb or damage the pipe. Work in and compact the haunching material in the area between the bedding and the underside of the pipe before placing and compacting the remainder of the embedment (see Figure 6-5). Do not permit compaction equipment to contact and damage the pipe. Use compaction equipment and techniques that are compatible with materials used and located in the trench.

Compaction of soils containing few fines (Class I and Class II with less than 5 percent fines). In-place densities of these materials are typically difficult to measure; therefore, the installation should be monitored to ensure that uncompacted materials are worked in to completely fill voids in the haunch zone. If compaction is required, use surface plate vibrators, vibratory rollers, or internal vibrators. The compacted lift thickness should not exceed 12 in. (300 mm) when compacted with surface plate vibrators or vibratory rollers; when compacted with internal vibrators, it should not exceed the length of the internal vibrators. Density determination should typically be in accordance with ASTM D4253 or ASTM D7382. In some cases, laboratory densities of SW or SP soils may be determined by ASTM D698 if a clearly defined compaction curve can be established.

Compaction of soils containing some fines (Class II with 5–12 percent fines). The method for compacting these soils should be based on fine content. The methods of

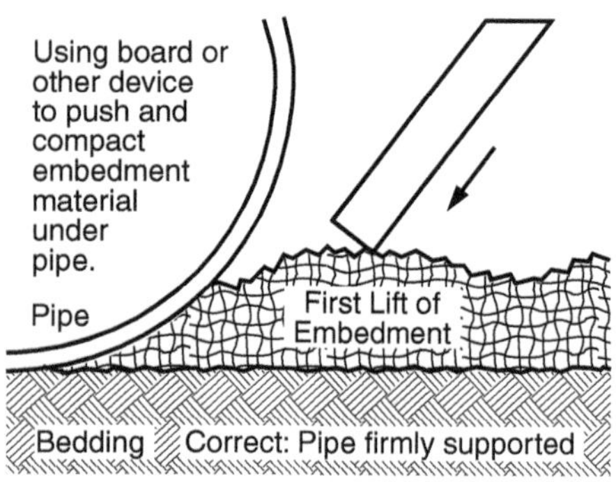

a. Ensuring Firm Pipe Support

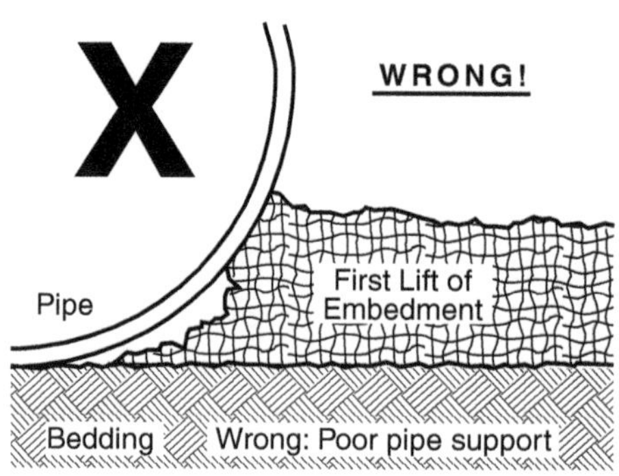

b. Improper Haunch

Source: Flowtite Technology, Sandefjord, Norway

Figure 6-5 Proper compaction under haunches

compaction and density determination should be based on the method that results in the higher in-place density. Use vibratory methods for soils with lower fine content and impact methods for soils with higher fine content. Laboratory density determination should typically be in accordance with ASTM D4253, ASTM D7382, or ASTM D698.

Compaction of soils containing a significant amount of fines (Class III, Class IV, and Class V). These soils should be compacted with impact tampers. Laboratory density determination should be in accordance with ASTM D698. The maximum density occurs at the optimum moisture content. Less effort is required to reach a given density when the moisture content is within 2 percentage points of the optimum moisture. A rapid method of determining the PC and moisture variation is described in ASTM D5080. For compaction levels of 90 percent or higher, the compacted lift thickness should not exceed 6 in. (150 mm).

Determination of the in-place density of soils. The in-place density of any in situ or fill soil may be determined in accordance with ASTM D1556, ASTM D2167, ASTM D4914, or ASTM D5030. The applicable test method will depend on the type of soil, moisture content of the soil, and the maximum particle size present in the soil. The moisture content of the soil may be determined in accordance with ASTM D2216, ASTM D4643, ASTM D4944, or ASTM D4959. When using nuclear density-moisture gauges (ASTM D6938), the gauge should be site-calibrated in the proximity of the pipe and in the excavation as recommended by the gauge manufacturer.

Minimum density. The minimum embedment density should be established based on an evaluation of specific project conditions. Densities higher than those recommended in Table 6-2 may be appropriate. Minimum densities given in Table 6-2 are intended to provide satisfactory embedment stiffness in most installation conditions.

Densification using water. Densification of embedment using water (jetting or saturation with vibration) should be done only under controlled conditions.

Backfill around angularly deflected pipe joints. When pipe joints are angularly rotated to accomplish large radii bends in pipelines that will operate at internal pressures of 15 psi (100 kPa) or greater, the joint should be embedded with compacted Class I materials, 90 percent compacted Class II materials, or 95 percent compacted Class III or Class IV materials.

Consult the manufacturer for minimum depths of burial and additional restraint that may be required when the angular deflection is vertical.

Minimum cover. To preclude damage to the pipe and disturbance of embedment, a minimum depth of backfill above the pipe should be maintained before allowing vehicles or heavy construction equipment to traverse the pipe trench. The minimum depth of cover should be established based on an evaluation of specific project conditions, such as pipe diameter and stiffness, soil type and stiffness, and live load type and magnitude. In the absence of an engineering evaluation, the following minimum cover requirements should be used.

For embedment materials installed to the minimum densities given in Table 6-2 and live loads similar to AASHTO H-20, provide cover (i.e., depth of backfill above top of pipe) of at least 24 in. (0.6 m) for Class I embedment and at least 36 in. (0.9 m) for Class II, Class III, or Class IV embedment before allowing vehicles or construction equipment to traverse the trench surface; provide at least 48 in. (1.2 m) of cover before using a hydro hammer for compaction. Where construction loads may be excessive (e.g., cranes, earth-moving equipment, or other vehicles with wheel loads that exceed the AASHTO H-20 loading), minimum cover should be increased or special structures such as relief slabs at grade installed to reduce the load transferred to the pipe.

If a risk of pipe flotation occurs, the minimum cover should be one pipe diameter. If a specific analysis is made of the buoyant force of an empty pipe compared with the submerged weight of soil over the pipe, this minimum cover may be reduced.

6.7.4 Connections and Appurtenant Structures

Connections to manholes and rigid structures and changing foundation soils. When differential settlement can be expected—such as at the ends of casing pipe, when the pipe enters a manhole, at anchor blocks, or where foundation soils change stiffness—a flexible system capable of accommodating the anticipated settlement should be provided. This may be accomplished by placing a joint as close as practically possible to the face of the structure and a second joint within one to two pipe diameters of the face of the structure (see Figure 6-3). The short length of pipe, called a "rocker pipe," shall be installed in straight alignment with the short pipe section coming out of the structure. The rocker pipe should have a minimum pipe stiffness of 36 psi (248 kPa) to transition between lower-stiffness pipe and the rigid structure. Multiple rocker pipes should not be used. Alternatively, attach the pipe to the rigid structure with a flexible boot capable of accommodating the anticipated differential movement. Extra care and caution must be taken to replace and properly compact backfill adjacent to any rigid structure. Construction of concrete structures will frequently require overexcavation for formwork and so on. This extra-excavated material must be restored to a density level compatible with surroundings to prevent excess deformation and/or joint rotation adjacent to the structure. In these areas, backfill should be compacted to achieve the same soil density as specified for all pipe backfill but not less than required to achieve a soil modulus (M_{sb}) of at least 1,000 psi (6.9 MPa). The use of cement-stabilized backfills adjacent to large structures has been found to be effective in preventing excess deformation where diameters are larger than about 60 in. (1,500 mm). Other methods of accommodating the differential settlements may be acceptable.

Vertical risers. Provide support for vertical risers as commonly found at service connections, cleanouts, and drop manholes to preclude vertical or lateral movement. Prevent the direct transfer of thrust as a result of surface loads and settlement and ensure adequate support at points of connection to main lines.

Exposing pipe for making service line connections. When excavating for a service line connection, material should be excavated from above the top of the existing pipe before removing material from the sides of the pipe. When backfilling excavations of existing lines, the materials and construction methods used should restore the installation to its condition before excavation.

Pipe caps and plugs. Secure caps and plugs to the pipe to prevent movement and resulting leakage under test and service pressures. If lines are to be tested under pressure, any plugs and caps must be designed to safely carry the test pressure.

Parallel piping systems. Compact the soil between the pipes in the same manner as when compacting the soil between the pipe and the trench wall, taking special care to compact the soil in the haunches.

6.7.5 Thrust Blocks

When restrained types of joints are used for assembly of buried fiberglass-reinforced plastic piping, thrust blocks may not be necessary. This shall be validated by system stress analysis and appropriate qualification testing of the components. Installation requirements related to thrust blocks are discussed in Chapter 7.

6.8 FIELD MONITORING

Compliance with installation requirements for trench depth, grade, water conditions, foundation, embedment and backfill materials, joints, density of materials in place, and safety should be monitored to ensure conformance with the contract documents.

Deflection. Monitor the deflection level in the pipe throughout the installation process for conformance to the requirements of the contract specifications and the manufacturer's recommendations. Conduct deflection measurement programs early in a project to verify that the construction procedures being used are adequate. The allowable deflection at the time of installation is the long-term allowable deflection reduced by the effects of deflection lag. If necessary, also consider the effects of vertical ovaling during compaction. Complete all deflection checks before conducting any pressure tests.

Pressure testing. Most pressure pipelines are tested after installation to detect leaks, installation flaws, damaged pipes, or other deficiencies. As a general rule, such tests should not be conducted using air pressure unless special precautions, not within the scope of this manual, are used. Additional recommendations for conducting pressure tests include the following.

- Required thrust restraints are properly installed (and sufficiently cured if applicable).
- Backfilling should be completed. Some sections of the line may be left uncovered, provided suitable lateral and longitudinal restraint is provided.
- Pumps and valves are anchored.
- Test caps and end plugs are properly installed and restrained as necessary.
- The pipeline is vented while being filled to allow all air to escape.
- The line is slowly pressurized to avoid pressure surges.
- In determining the test pressure, the lowest point on the line will have the highest pressure. If the test pressure gauge is not installed at this location, the correct pressure should be determined by calculation.
- The test fluid temperature should be stable during the test period (to avoid pressure changes caused by thermal expansion or contraction that may be misinterpreted as leaks).
- Typical maximum field test pressure of the installed pipe is $1.5 \times P_c$ unless the pipe is specifically designed for a higher value.

6.9 CONTRACT DOCUMENT RECOMMENDATIONS

The following guidelines may be included in contract documents for a specific project to cover installation requirements; ASTM D3839 provides similar guidelines. In either case, applications for a particular project may require that the engineer provide more specific requirements in several areas, including

- maximum particle size if different from Sec. 6.5.2;
- restrictions on use of classes of embedment and backfill materials;
- specific gradations of embedment materials for resistance to migration;
- state-specific restrictions on leaving trenches open;
- restrictions on mode of dewatering and design of underdrains;
- requirements on minimum trench width;
- restrictions or details for support of trench walls;
- specific bedding and foundation requirements;
- specific restrictions on methods of compaction;
- minimum embedment density if different from these recommendations (specific density requirements for backfill [e.g., for pavement subgrade]);

- minimum cover requirements;
- detailed requirements for support of vertical risers, standpipes, and stacks to accommodate anticipated relative movements between pipe and appurtenances;
- detailing to accommodate thermal movements, particularly at risers;
- detailed requirements for manhole connections;
- requirements on methods of testing compaction and leakage; and
- requirements on deflection and deflection measurements, including method and time of testing.

6.10 REFERENCES

AASHTO (American Association of State Highway and Transportation Officials). 2020 (9th ed). *LRFD Bridge Design Specifications*. Washington, D.C.: AASHTO.

AASHTO (American Association of State Highway and Transportation Officials). 1991. M 145, Standard Specification for Classification of Soils and Soil-Aggregate Mixtures for Highway Construction Purposes. Washington, D.C.: AASHTO.

ANSI/AWWA (American National Standards Institute/American Water Works Association). 2020. C950. Fiberglass Pressure Pipe. Denver: AWWA.

Chapter 7

Buried Pipe Thrust Restraints

7.1 UNBALANCED THRUST FORCES

Unbalanced thrust forces occur in pressure pipelines at changes in direction (i.e., elbows, wyes, tees, etc.), at changes in cross-sectional area (i.e., reducers), or at pipeline terminations (i.e., bulkheads). These forces, if not adequately restrained, may cause pipeline movement resulting in separated joints and/or pipe damage. Thrust forces are as follows: (1) hydrostatic thrust due to internal pressure of the pipeline; and (2) hydrodynamic thrust due to changing momentum of flowing fluid. Because most pressure lines operate at relatively low velocities, the hydrodynamic force is very small and usually ignored. Thrust blocks will be required for systems with unrestrained joints. When restrained-type joints are used for assembly of buried fiberglass-reinforced plastics piping, thrust blocks may not be necessary. This will be validated by system stress analysis and appropriate qualification testing of the components.

The equations in this chapter are presented with inch-pound units in the left column and metric units in the right column.

7.1.1 Hydrostatic Thrust

Typical examples of hydrostatic thrust are shown in Figure 7-1. The thrust in dead ends, tees, laterals, and reducers is a function of internal pressure P and cross-sectional area A at the pipe joint. The resultant thrust at a bend is also a function of the deflection angle Δ and is given by

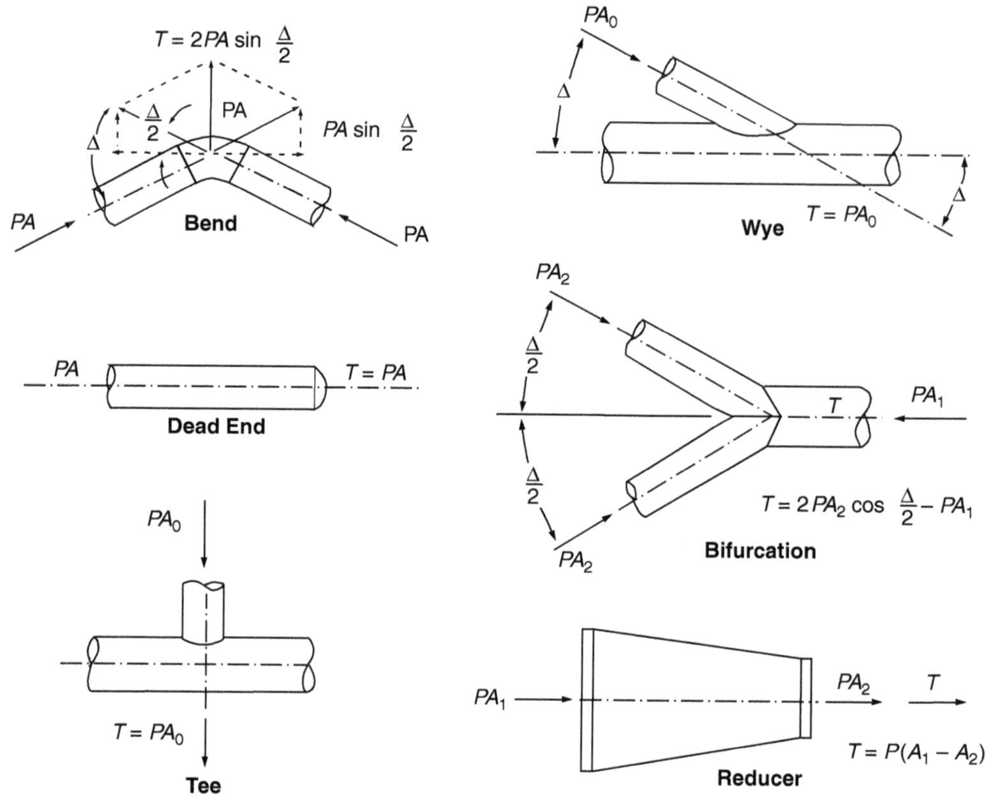

Figure 7-1 Thrust force definitions

$$T = 2PA \sin(\Delta/2) \qquad T = 2{,}000PA \sin(\Delta/2) \qquad \text{(Eq 7-1)}$$

Where:
- T = hydrostatic thrust, lb
- P = internal pressure, psi
- A = cross-sectional area of pipe joint, in.^{2}
  = $(\pi/4)\,D_j^2$
- D_j = joint diameter, in.
- Δ = bend angle, degrees

Where:
- T = hydrostatic thrust, N
- P = internal pressure, kPa
- A = cross-sectional area of pipe joint, m^2
  = $(\pi/4)\,(D_j/1{,}000)^2$
- D_j = joint diameter, mm
- Δ = bend angle, degrees

7.2 THRUST RESISTANCE

For buried pipelines, unbalanced horizontal thrust forces have two inherent sources of resistance: (1) frictional drag from dead weight of the pipe, earth cover, and contained fluid; and (2) passive resistance of soil against the pipe or fitting in the direction of the thrust. If this resistance is not sufficient to resist the thrust, it must be supplemented by increasing the supporting area on the bearing side of the fitting with a thrust block, increasing the frictional drag of the line by "tying" adjacent pipe to the fitting, or otherwise anchoring the fitting to limit or prevent movement. Unbalanced uplift thrust at a vertical deflection is resisted by the dead weight of the fitting, earth cover, and contained fluid. If this type of resistance is not sufficient to resist the thrust, it must be supplemented by increasing the dead weight with a gravity-type thrust block, increasing the dead weight of the line by "tying" adjacent pipe to the fitting, or otherwise anchoring the fitting to limit or prevent movement.

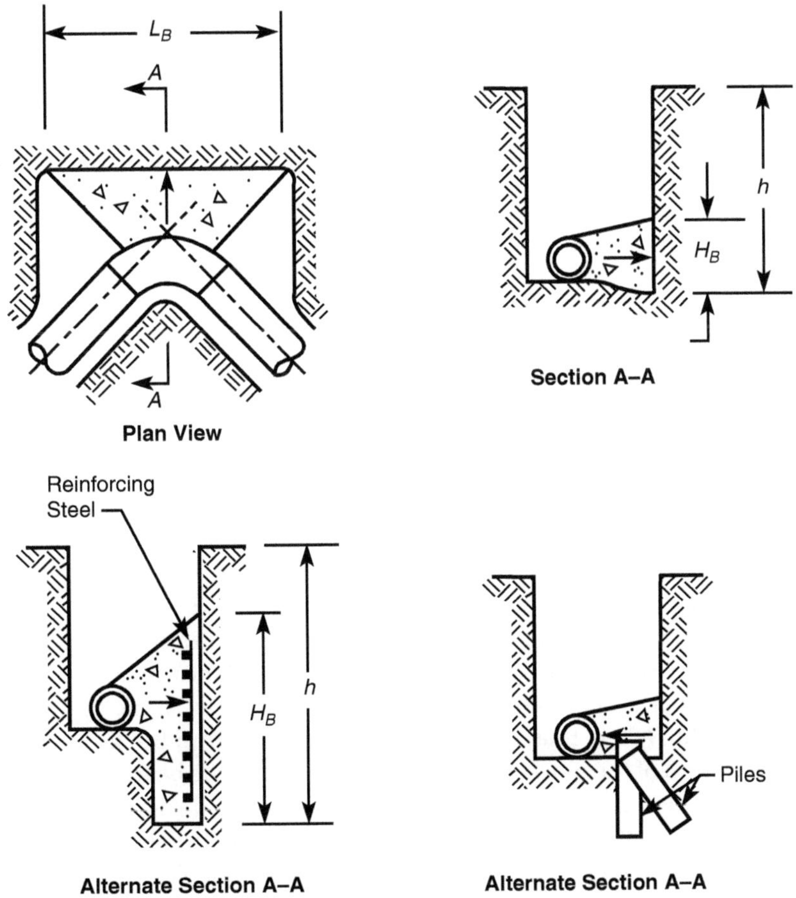

Figure 7-2 Typical thrust blocking of a horizontal bend

7.3 THRUST BLOCKS

Concrete thrust blocks increase the ability of fittings to resist movement by increasing the bearing area and the dead weight of the fitting. Typical thrust blocking of a horizontal bend (elbow) is shown in Figure 7-2.

7.3.1 Calculation of Size

Ignoring the dead weight of the thrust block, the block size can be calculated based on the bearing capacity of the soil:

$$\text{Area of block} = (L_B)(H_B) = (T \times FS)/\sigma$$

Where:

$(L_B)(H_B)$ = area of bearing surface of thrust block, ft^2

T = thrust force, lb

FS = design factor, 1.5

σ = bearing strength of soil, lb/ft^2

$$\text{Area of block} = (L_B)(H_B) = (T \times FS)/1{,}000\ \sigma \quad \text{(Eq 7-2)}$$

Where:

$(L_B)(H_B)$ = area of bearing surface of thrust block, m^2

T = thrust force, N

FS = design factor, 1.5

σ = bearing strength of soil, kPa

Table 7-1 Horizontal soil-bearing strengths

Soil	Bearing Strength	
	*σ lb/ft²**	*kN/m²*
Muck	0	0
Soft clay	1,000	48
Silt	1,500	72
Sandy silt	3,000	144
Sand	4,000	192
Sandy clay	6,000	287
Hard clay	9,000	431

* Although the bearing strength values have been used successfully in the design of thrust blocks and are considered conservative, their accuracy is dependent on accurate soil identification and evaluation. The design engineer must select the proper bearing strength of a particular soil type.

Typical values for conservative horizontal bearing strengths of various soil types are listed in Table 7-1. If it is impractical to design the block for the thrust force to pass through the geometric center of the soil-bearing area, the design should be evaluated for stability.

After calculating the concrete thrust block size and reinforcement, if necessary, based on the bearing capacity of soil, the shear resistance of the passive soil wedge behind the thrust block should be checked because it may govern the design. For a thrust block having its height H_B less than one-half of the distance from the ground surface to base of block h, the design of the block is generally governed by the bearing capacity of the soil. However, if the height of the block H_B exceeds one-half h, the design of the block is generally governed by shear resistance of the soil wedge behind the thrust block. Determining the value of the bearing and shear resistance of the soil and thrust block reinforcement is beyond the scope of this manual. For more detailed information, a qualified geotechnical engineer should be consulted.

7.3.1.1 Typical configurations. Determining the bearing value σ is the key to "sizing" a thrust block. Values can vary from less than 1,000 lb/ft^2 (48 kN/m^2) for very soft soils to several tons per square foot (kilonewtons per meters squared) for solid rock. Knowledge of local soil conditions is necessary for proper sizing of thrust blocks. Figure 7-2 shows several details for distributing thrust at a horizontal bend. Section A-A is the more common detail, but the other methods shown in the alternate sections may be necessary in weaker soils. Figure 7-3 illustrates typical thrust blocking of vertical bends. Design of the block for a bottom bend is the same as for a horizontal bend, but the block for a top bend must be sized to adequately resist the vertical component of thrust with dead weight of the block, bend, water in the bend, and overburden.

7.3.1.2 Proper construction is essential. Most thrust block failures can be attributed to improper construction. Even a correctly sized block can fail if it is not properly constructed. A block must be placed against undisturbed soil, and the face of the block must be perpendicular to the direction of and centered on the line of action of the thrust. A surprising number of thrust blocks fail because of inadequate design or improper construction. Many people involved in construction and design do not realize the magnitude of the thrusts involved. As an example, a thrust block behind a 36-in. (900-mm) 90-degree bend operating at 100 psi (689 kPa) must resist a thrust force in excess of 150,000 lb (667 kN). Another factor frequently overlooked is that thrust increases in proportion to the square of pipe diameter. A 36-in. (900-mm) pipe produces approximately four times the thrust produced by an 18-in. (450-mm) pipe operating at the same internal pressure.

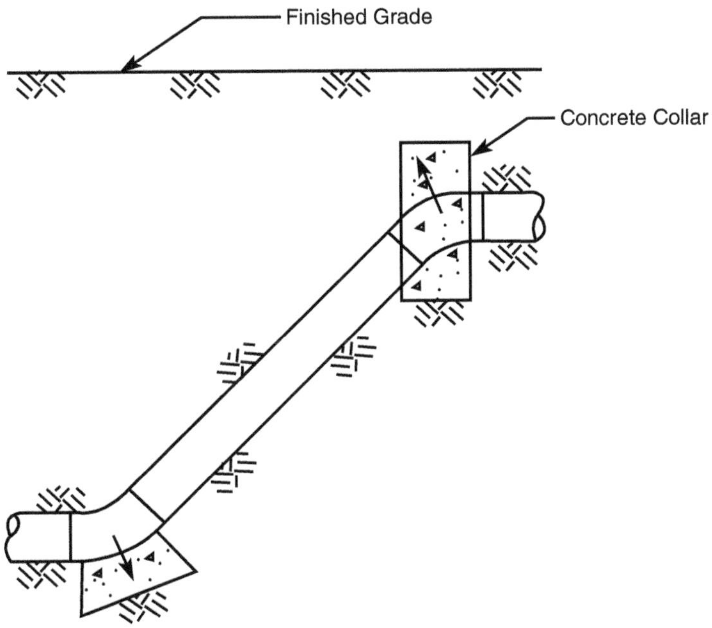

Figure 7-3 Typical profile of vertical bend thrust blocking

7.3.2 Adjacent Excavation

Even a properly designed and constructed thrust block can fail if the soil behind the block is disturbed. Properly sized thrust blocks have been poured against undisturbed soil only to fail because another utility or an excavation immediately behind the block collapsed when the line was pressurized. If the risk of future nearby excavation is high, the use of restrained (tied) joints may be appropriate.

7.4 JOINTS WITH SMALL DEFLECTIONS

The thrust at pipe joints installed with angular deflection is usually so small that supplemental restraint is not required.

7.4.1 Small Horizontal Deflections

Thrust T at horizontal deflected joints is resisted by friction on the top and bottom of the pipe, as shown in Figure 7-4. Additional restraint is not required when

$$T \leq fL_p\left(W_p + W_w + 2W_e\right)$$

$$T \leq fL_p\left(W_p + W_w + 2W_e\right) \quad \text{(Eq 7-3)}$$

Where:

T = $2PA \sin(\theta/2)$, lb
θ = angle of deflected joint, degrees
f = coefficient of friction
L_p = length of pipe, ft
W_p = weight of pipe, lb/LF
W_w = weight of fluid in pipe, lb/LF
W_e = earth cover load, lb/LF

Where:

T = $2{,}000PA \sin(\theta/2)$, N
θ = angle of deflected joint, degrees
f = coefficient of friction
L_p = length of pipe, m
W_p = weight of pipe, N/m
W_w = weight of fluid in pipe, N/m
W_e = earth cover load, N/m

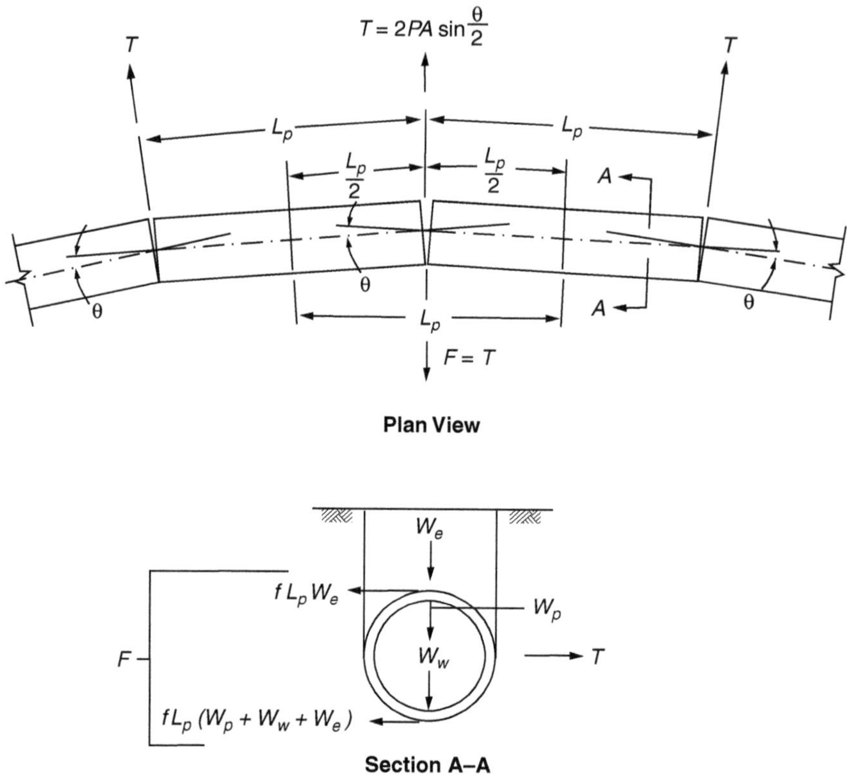

Figure 7-4 Restraint of thrust at deflected joints on long-radius horizontal curves

The passive soil resistance of the trench backfill against the pipe is ignored in the previous analysis. Depending on the installation and field conditions, the passive soil resistance of the backfill may be included to resist thrust.

The selection of a value for the coefficient of friction f is dependent on the type of soil and the roughness of the pipe exterior. Design values for the coefficient of friction generally vary from 0.25 to 0.50.

Determination of earth cover load should be based on a backfill density and height of cover consistent with what can be expected when the line is pressurized. Values of soil density vary from 90 to 130 lb/ft^3 (14 to 20 kN/m^3), depending on the degree of compaction (see Figure 7-5). W_e may be conservatively determined using the Marston equation for loads imparted to a flexible pipe, as follows:

$$W_e = (C_d)(W)(B_d)(B_c) \qquad W_e = (C_d)(W)(B_d)(B_c) \qquad \text{(Eq 7-4)}$$

Where:

W_e = earth load, lb/LF of pipe length
C_d = coefficient based on soil and the ratio of H and B_d (see Figure 7-5)
H = depth of fill to pipe top, ft
W = unit weight of soil, lb/ft^3
B_d = trench width at pipe top, ft
B_c = OD of pipe, ft

Where:

W_e = earth load, N/m of pipe length
C_d = coefficient-based soil and the ratio of H and B_d (see Figure 7-5)
H = depth of fill to pipe top, m
W = unit weight of soil, N/m^3
B_d = trench width at pipe top, m
B_c = OD of pipe, m

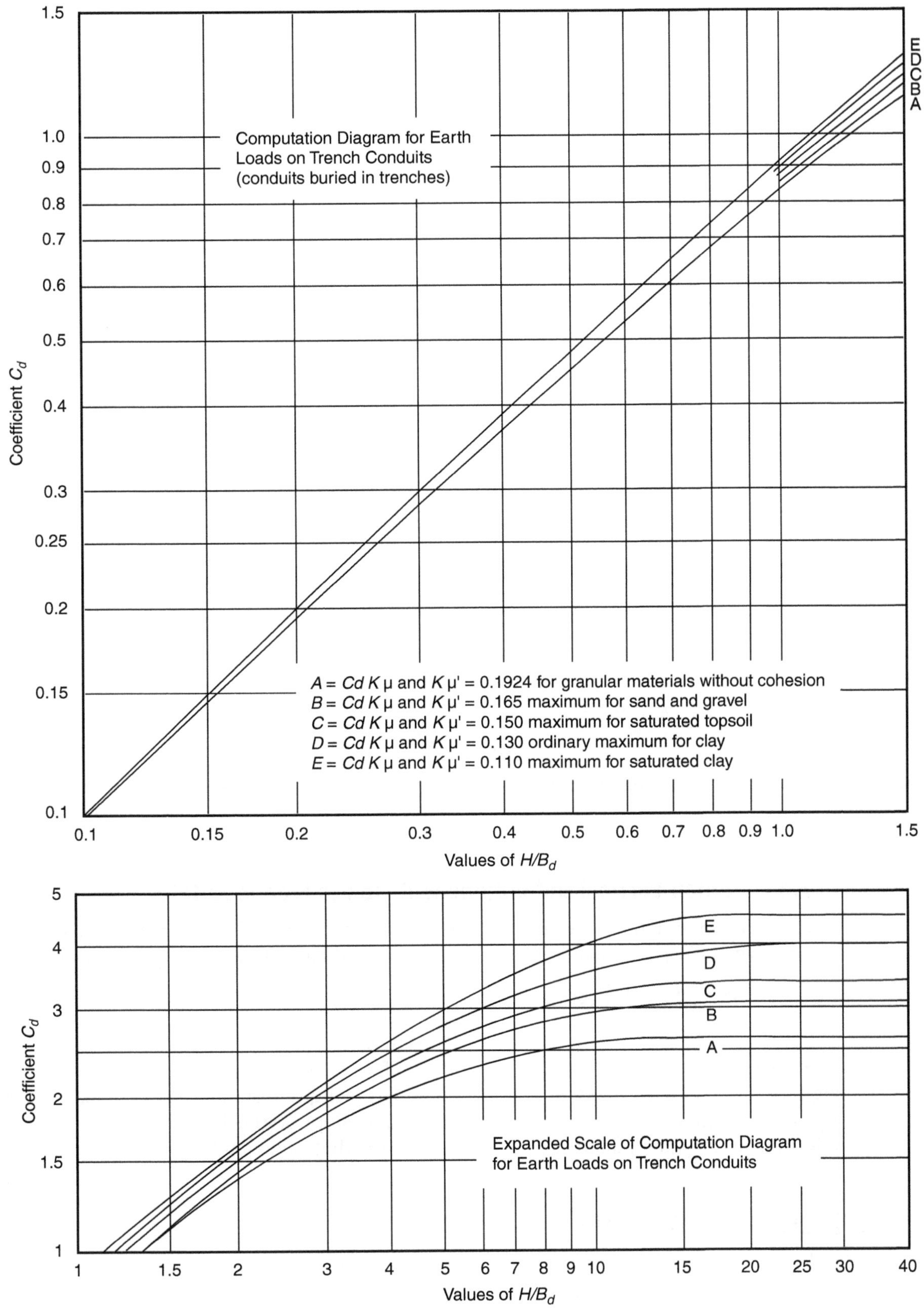

Figure 7-5 Computation diagram for earth loads on trench conduits

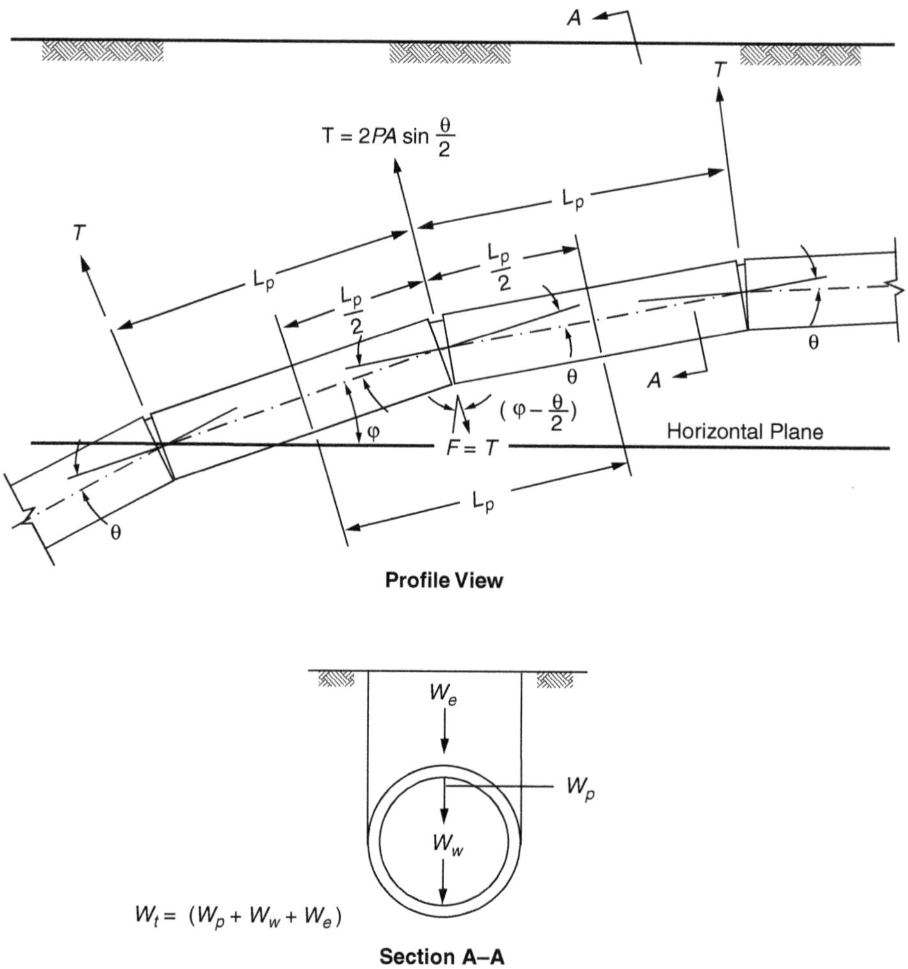

Figure 7-6 Restraint of uplift thrust at deflected joints on long-radius vertical curves

7.4.2 Small Vertical Deflections with Joints Free to Rotate

Uplift thrust at deflected joints on long-radius vertical curves is resisted by the combined dead weight W_t, as shown in Figure 7-6. Additional restraint is not required when

$$T \le L_p\left(W_p + W_w + W_e\right)\cos(\varphi - \theta/2) \qquad T \le L_p\left(W_p + W_w + W_e\right)\cos(\varphi - \theta/2) \qquad \text{(Eq 7-5)}$$

Where:

T = $2PA \sin(\theta/2)$, lb
L_p = length of standard or beveled pipe, ft
φ = slope angle, degrees
θ = deflection angle of joint, degrees

Where:

T = $2{,}000PA \sin(\theta/2)$, N
L_p = length of standard or beveled pipe, m
φ = slope angle, degrees
θ = deflection angle of joint, degrees

7.5 RESTRAINED (TIED) JOINTS

Unbalanced thrust forces at fittings or deflected joints may be resisted by using restrained joint(s) across the deflected joint or by tying adjacent pipes to the fitting. This method fastens several pipes on each side of the fitting to increase the frictional drag of the connected pipe to resist the fitting thrust. Because thrust diminishes from a maximum value at a fitting to zero at distance L from the fitting, requirements for longitudinal strength to resist thrust can be calculated for the pipe length immediately adjacent to the fitting and prorated on a straight-line basis for the remainder of the pipe within the tied distance L. Frictional resistance on the tied pipe acts in the opposite direction of resultant thrust T. Section A-A in Figure 7-4 shows the external vertical forces acting on a buried pipe with horizontal thrust and the corresponding frictional resistance. Uplift thrust restraint provided by gravity-type thrust blocks, shown for the top bend in Figure 7-3, may also be provided by the alternate method of increasing the dead weight of the line by tying adjacent pipe to the vertical bend. Section A-A in Figure 7-6 shows the vertical forces acting on a buried vertical (uplift) bend used in determining the thrust resistance by dead weight.

As previously stated, both analyses ignore the passive soil resistance of the backfill against the pipe. Depending on the installation and field conditions, the passive soil resistance of the backfill may be included to resist thrust.

7.5.1 Horizontal Bends and Bulkheads

As illustrated in Figure 7-7, the frictional resistance F needed along each leg of a horizontal bend is $PA \sin(\Delta/2)$. Frictional resistance per linear foot of pipe against soil is equal to

$$F = f\left(2W_e + W_p + W_w\right) \qquad F = f\left(2W_e + W_p + W_w\right) \qquad \text{(Eq 7-6)}$$

Where:

f = coefficient of friction between pipe and soil

F = frictional resistance, lb/ft

Where:

f = coefficient of friction between pipe and soil

F = frictional resistance, N/m

Therefore, the length of pipe L to be tied to each leg of a bend is calculated as

$$L_{bend} = \frac{1{,}000PA\sin(\Delta/2)}{f\left(2W_e + W_p + W_w\right)} \qquad L_{bend} = \frac{1{,}000PA\sin(\Delta/2)}{f\left(2W_e + W_p + W_w\right)} \qquad \text{(Eq 7-7)}$$

Where:

L_{bend} = length of pipe tied to each bend leg, ft

Where:

L_{bend} = length of pipe tied to each bend leg, m

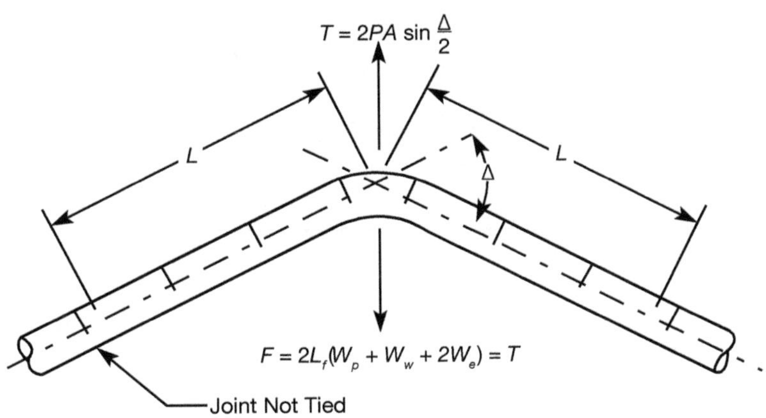

Figure 7-7 Thrust restraint with tied joints at bends

The length of pipe to be tied to a bulkhead or tee leg is

$$L_{bulk} = \frac{PA}{f\left(2W_e + W_p + W_w\right)} \qquad L_{bulk} = \frac{PA}{f\left(2W_e + W_p + W_w\right)} \qquad \text{(Eq 7-8)}$$

Where:
L_{bulk} = length of pipe tied to bulkhead or tee lag, ft, with all other variables as defined previously

Where:
L_{bulk} = length of pipe tied to bulkhead or tee lag, m, with all other variables as defined previously

7.5.2 Vertical (Uplift) Bends

As illustrated in Figure 7-8, the dead weight resistance needed along each leg of a vertical bend is 2*PA* sin (Δ/2). Dead weight resistance per linear foot of pipe in a direction opposite to thrust is

$$D_{wr} = \left(W_e + W_p + W_w\right) \cos\left(\varphi - \Delta/2\right) \qquad D_{wr} = \left(W_e + W_p + W_w\right) \cos\left(\varphi - \Delta/2\right) \qquad \text{(Eq 7-9)}$$

Where:
D_{wr} = dead weight resistance, lb/ft
φ = slope angle, degrees (see Figure 7-8)
Δ = bend angle, degrees (see Figure 7-8)

Where:
D_{wr} = dead weight resistance, N/m
φ = slope angle, degrees (see Figure 7-8)
Δ = bend angle, degrees (see Figure 7-8)

Length of pipe *L* to be tied to leg of a vertical (uplift) bend is calculated as

$$L_t = \frac{PA\sin\left(\Delta/2\right)}{\left(W_e + W_p + W_w\right)\cos\left[\varphi - \left(\Delta/2\right)\right]}. \qquad L_t = \frac{1{,}000PA\sin\left(\Delta/2\right)}{\left(W_e + W_p + W_w\right)\cos\left[\varphi - \left(\Delta/2\right)\right]} \qquad \text{(Eq 7-10)}$$

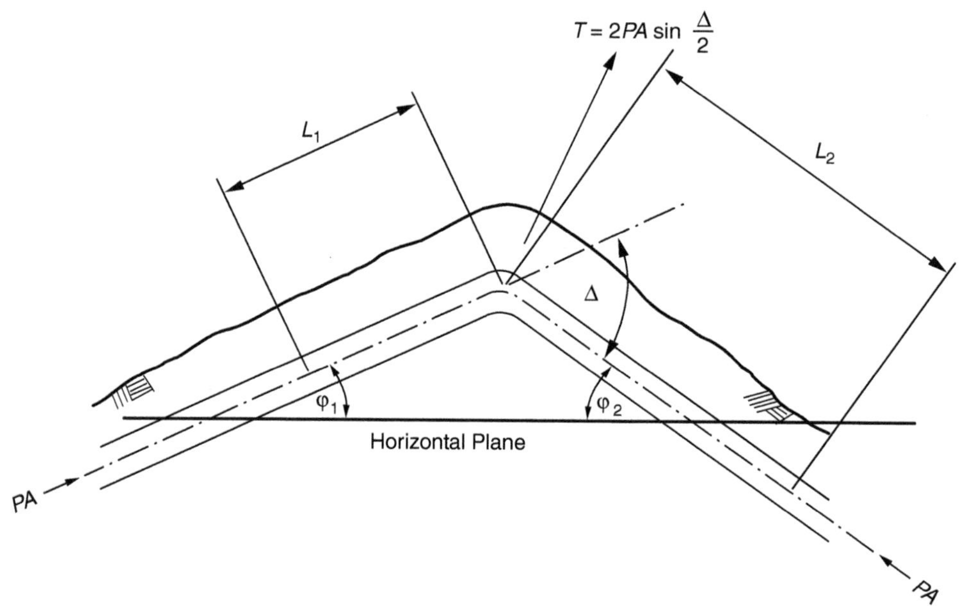

Figure 7-8 Length of tied pipe on each leg of vertical (uplift) bend

Where:

L_t = length of pipe to be tied to one of the legs of a vertical (uplift) bend, ft (see Figure 7-8 and note that angles φ_1 and φ_2 may be equal or different).

Where:

L_t = length of pipe to be tied to one of vertical (uplift) bend, m (see Figure 7-8 and note that angles φ_1 and φ_2 may be equal or different).

Vertical downward bends are resisted by bearing of the trench against the bottom of the pipe. Properly bedded pipe should not have to be investigated for this condition.

7.5.3 Transmission of Thrust Force Through Pipe

In addition to calculating pipe length to be tied to a fitting, engineers must be sure that tied pipe lengths have sufficient strength in the longitudinal direction to transmit thrust forces. The maximum thrust force for which the pipe adjacent to a bend must be designed is equal to

$$F_y = 0.001\left(5.43\Delta + 0.063\Delta^2\right)PA \qquad F_y = \left(5.43\Delta + 0.063\Delta^2\right)PA \qquad \text{(Eq 7-11)}$$

Where:

F_y = maximum axial thrust force for which the pipe adjacent to a bend must be designed, lb

NOTE: For $\Delta = 90°$; $F_y = PA$

Where:

F_y = maximum axial thrust force for which the pipe adjacent to a bend must be designed, N

NOTE: For $\Delta = 90°$; $F_y = PA$

This page intentionally blank.

Chapter 8

Aboveground Pipe Design and Installation

8.1 INTRODUCTION

This chapter addresses the design and installation of fiberglass pipeline systems in aboveground applications for pipe diameters of 16 in. (400 mm) and smaller and only for pipelines that have restrained joints. Different design provisions and supporting methods may be applicable for specific project requirements, larger diameters, or a particular piping product. Consult with the manufacturer and the piping engineer for appropriate design considerations.

The equations in this chapter are presented with inch-pound units in the left column and metric units in the right column.

8.2 THERMAL EXPANSION AND CONTRACTION

Fiberglass pipe may have a different expansion rate in the hoop and axial directions. For example, a filament-wound pipe with a 55° winding angle has approximately the same thermal expansion as steel in the hoop direction and approximately twice the thermal expansion as steel in the axial direction. The total expansion or contraction for a pipe system is determined using the following equation:

$$L_c = (12)(C_t)(L)(T_c) \qquad L_c = (1{,}000)(C_t)(L)(T_c) \qquad \text{(Eq 8-1)}$$

Where:

L_c = length change, in.

C_t = coefficient of axial thermal expansion in./in./°F

L = length between anchors, ft

T_c = temperature change, °F (maximum operating temperature minus installation temperature for expansion and installation temperature minus minimum operating temperature for contraction)

Where:

L_c = length change, mm

C_t = coefficient of axial thermal expansion mm/mm/°C

L = length between anchors, m

T_c = temperature change, °C (maximum operating temperature minus installation temperature for expansion and installation temperature minus minimum operating temperature for contraction)

8.3 THERMAL EXPANSION DESIGN

In the design of aboveground pipelines, the supports and guides for the pipe become important considerations because of thermal expansion.

In addition to pressure resistance and life limitations, the effects of thermal expansion and contraction should be considered. Many methods accommodate the length changes associated with thermal expansion and contraction. The four most commonly used methods include

- anchoring and guiding,
- direction changes,
- expansion loops, and
- mechanical expansion joints.

Guides, expansion loops, and mechanical expansion joints are installed in straight lines and are anchored at each end. Experience has shown that direction changes are the least expensive method of accommodating thermal expansion. Guide spacing is the next most economical method, followed by mechanical expansion joints and expansion loops.

For small temperature changes and piping systems that consist of short run lengths, it is usually unnecessary to make special provisions for thermal expansion. However, any system should have the capability of accommodating length changes.

Experience has shown that aboveground piping systems need anchors at approximately 300-ft (90-m) intervals. (Note: This value may vary for larger pipe sizes.) These anchors limit pipe movement caused by vibrations and transient loading conditions. Anchors should fasten all transition points within the system. Transition points are places where pipe diameter, material, elevation, or direction changes or where the manufacturer changes. Anchors at transition points limit the transfer of thermal end loads from line section to line section.

8.3.1 Thermal End Loads

The axial modulus of elasticity of fiberglass pipe can vary from approximately 1.5 to 10 percent of steel. The low modulus results in lower end loads that require restraining equipment less strong than that used for metallic pipelines. Internal pressures in the piping system can result in some length change. Experience has shown that this elongation is often insignificant and may not need to be considered in the design.

The equation for calculating the thermal end load is

$$E_L = (C_t)(E)(A)(T_c) \qquad E_L = (C_t)(E)(A)(T_c) \times 10^9 \qquad \text{(Eq 8-2)}$$

Where:

E_L = thermal end load, lb
E = axial modulus, psi
A = cross-sectional area of pipe wall, in.^{2}
= $\pi/4(OD^2 - ID^2)$
OD = pipe outside diameter, in.
ID = pipe inside diameter, in.

Where:

E_L = thermal end load, N
E = axial modulus, GPa
A = cross-sectional area of pipe wall, m^2
= $\pi/4(OD^2 - ID^2) \times 10^{-6}$
OD = pipe outside diameter, mm
ID = pipe inside diameter, mm

When pipe between anchors expands, the pipe undergoes compression. When contraction occurs, the pipe experiences tension.

8.3.2 Spacing Design: Anchoring and Guiding

Installing anchors at all direction and elevation changes serves to divide the system into straight runs. With anchors installed, guides are an economical method for dealing with expansion. The relatively low modulus of fiberglass pipe allows it to absorb the thermal stresses as compressive stresses in the pipe wall. Compressive stresses from expansion may result in buckling, unless the pipe is constrained at close intervals to prevent columnar instability.

The equation to calculate maximum allowable guide spacing interval is

$$L_G = \sqrt{\frac{(\pi^2)(E_b)(I)}{144(C_t)(A)(T_c)(E_c)}} \qquad L_G = \sqrt{\frac{(\pi^2)(E_b)(I)}{(C_t)(A)(T_c)(E_c)}} \qquad \text{(Eq 8-3)}$$

Where:

L_G = maximum guide (support) spacing, ft
E_b = axial bending modulus, psi
I = moment of inertia, in.^{4}
$= \dfrac{\pi(OD^4 - ID^4)}{64}$
E_c = axial compressive modulus, psi

Where:

L_G = maximum guide (support) spacing, m
E_b = axial bending modulus, GPa
I = moment of inertia, mm^4
$= \dfrac{\pi(OD^4 - ID^4) \times 10^{-12}}{64}$
E_c = axial compressive modulus, GPa

Because the bending and compressive moduli are obtained from experimental data, the ratio E_b/E_c, using data representative of the minimum and maximum operating temperatures, should be calculated. The lower value of the two calculations will satisfy the interest of conservative design.

Compare guide intervals with the intervals for supports, then adjust guide spacing for a better match with support spacing. For example, adjust intervals so a guide replaces every second or third support. Remember, all guides act as modified supports and must meet the minimum requirements for supports, anchors, and guides, as prescribed in other sections of this chapter.

8.3.3 Expansion Joint Design

Expansion joints may be used to absorb thermal expansion in long, straight pipe runs. Various types of expansion joints are available and suitable for use with fiberglass piping systems. Because the forces developed during a temperature change are relatively low compared with metallic systems, it is essential to specify an expansion joint that activates with low force. Fiberglass pipe will expand more than most metallic systems. The required movement per expansion joint and the number of expansion joints may be greater for fiberglass systems.

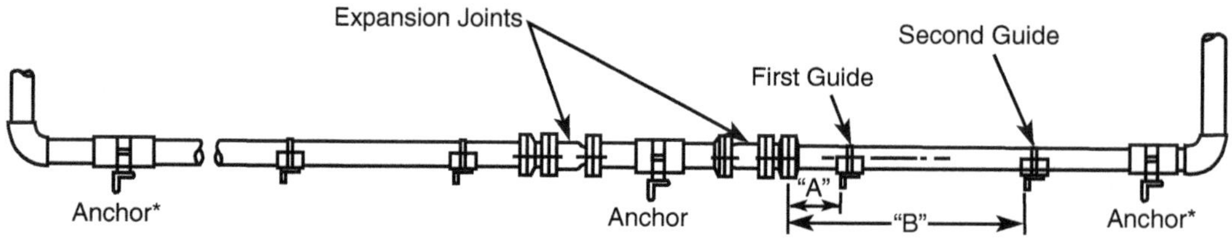

*Anchor Load = $\frac{\pi}{4}$ (ID)^2 × Internal Pressure

NOTE: A = 4 diameters; B = 14 diameters

Source: Reprinted with permission from Fiberglass Pipe Handbook, Fiberglass Pipe Institute, New York

Figure 8-1 Typical expansion joint installation

The allowable activation force for expansion joints depends on both the thermal forces developed in the pipe and the support or guide spacing. Guide spacing at the entry of an expansion joint is typically four pipe diameters (first guide) and 14 pipe diameters (second guide) from the inlet of the expansion joint (Figure 8-1). These guides and locations give proper alignment. The spacing of the remaining supports should remain within the maximum calculated interval.

The equation for calculating the allowable activation force is

$$P_{cr} = \frac{\pi^2 (E_c)(I)(S_f)}{{L_G}^2} \qquad P_{cr} = \frac{\pi^2 (E_c)(I)(S_f) \times 10^9}{{L_G}^2} \qquad \text{(Eq 8-4)}$$

Where:

P_{cr} = critical buckling force of pipe, lb

S_f = material variation safety factor; 0.9 recommended

Where:

P_{cr} = critical buckling force of pipe, N

S_f = material variation safety factor; 0.9 recommended

The pressure thrust must also be considered. Pressure thrust is the design pressure times the area of the expansion joint.

In all applications, the activation force of the expansion joint must not exceed the thermal end loads developed by the pipe. The cost and limited motion capability of expansion joints make them impractical to use in many applications. In these cases, loops, guide spacing, or short lengths of flexible hose can handle thermal expansion. The expansion joint needs an anchor on both sides for proper operation.

8.3.4 Expansion Loop Design

Expansion loops flex to accommodate changes in length (Figure 8-2). This design method is used to calculate the stress developed in a cantilevered beam with a concentrated load at the free end and ignores flexibility of the loop leg, the leg parallel to the line.

Two guides on both sides of each expansion loop ensure proper alignment. The recommended guide spacing is four (first guide) and 14 (second guide) nominal pipe diameters. Additional guides or supports should be located so the maximum spacing interval is not exceeded.

To design an expansion loop, use the following equation:

$$L_A = \sqrt{\frac{(K)(L_c)(E_b)(\text{OD})}{144(\sigma_b)}} \qquad L_A = \sqrt{\frac{(K)(L_c)(E_b)(\text{OD})}{1{,}000(\sigma_b)}} \qquad \text{(Eq 8-5)}$$

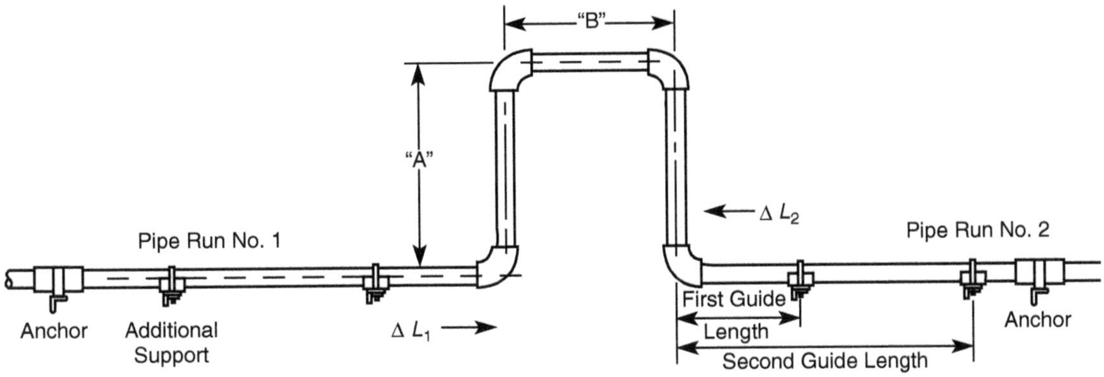

Source: Reprinted with permission from Fiberglass Pipe Handbook, Fiberglass Pipe Institute, New York

Figure 8-2 Expansion loop dimensions

Where:

L_A = length of the "A" leg, ft
K = cantilevered beam constant
= 0.75 for nonguided beam
= 3.0 for guided beam
σ_b = design-allowable bending stress, psi (A minimum safety factor of 8 on ultimate strength is recommended.)

Where:

L_A = length of the "A" leg, m
K = cantilevered beam constant
= 0.75 for nonguided beam
= 3.0 for guided beam
σ_b = design-allowable bending stress, MPa (A minimum safety factor of 8 on ultimate strength is recommended.)

If the maximum allowable bending stress of the fittings is greater than the maximum for the pipe, the bending moment of the fitting does not need to be considered. In other cases, the fitting manufacturer will provide allowable bending moments for the fittings. These values are used in Eq 8-6 to determine the A-leg length. The results are compared, and the larger value is used. Pipelines with heavy-wall pipe and relatively thin-wall fittings are most likely to require verification of the L_A dimension.

$$L_A = \sqrt{\frac{12(L_c)(E_b)(I)}{144(M)}} \qquad L_A = \sqrt{\frac{(L_c)(E_b)(I) \times 10^6}{(M)}} \qquad \text{(Eq 8-6)}$$

Where:

L_c = maximum length change (see Eq 8-1), in.
M = allowable elbow bending moment, lb-in.

Where:

L_c = maximum length change (see Eq 8-1), mm
M = allowable elbow bending moment, N-m

In some cases, the manufacturer may require anchors at all fittings. For example, mitered fittings and/or large-diameter fittings may have allowable bending stresses less than that of the pipe. In these cases, thermal expansion procedures may be limited to the use of anchors and guides or expansion joints if the bending moment is not available.

8.3.5 Direction Changes

In some installations, system directional changes can perform the same function as expansion loops. Directional changes that involve some types of fittings, such as saddles, should not be used to absorb expansion or contraction. The bending stresses may cause fitting failure. Stress in the pipe at a given directional change depends on the total change in length and the distance to the first secure hanger or guide past the directional change. In other words, the required flexible leg length is based on the maximum change in length.

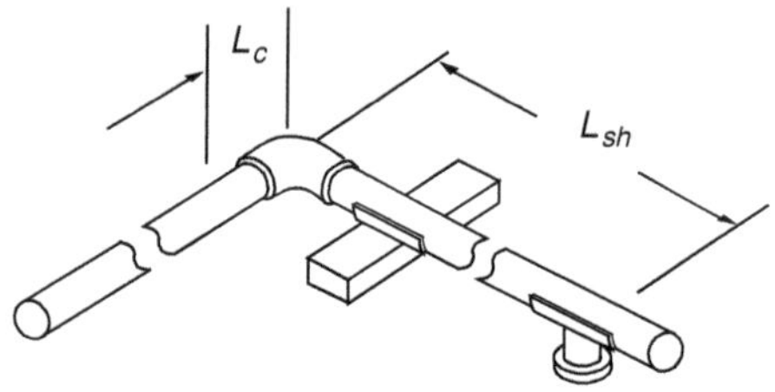

Figure 8-3 Directional change

Recommended support or guide spacing cannot be disregarded. However, flexible or movable supports, such as strap hangers, can provide support while allowing the pipe to move and absorb the changes in length. Supports must prevent lateral movement or pipe buckling.

Where large thermal movements are expected, a short length of flexible hose or flexible pipe installed at a change in direction will absorb some of the line movement. This method of handling thermal expansion is usually the most economical means of compensating for large displacements when the guide spacing method cannot be used. Hose or pipe manufacturers provide specifications giving the minimum bend radius, chemical compatibility, temperature, and pressure rating of a particular flexible hose or flexible pipe.

The equation for calculating the length of the flexible pipe leg (i.e., the distance to the first restraining support or guide) is

$$L_{sh} = \sqrt{\frac{1.5(L_c)(E_b)(\text{OD})}{144(\sigma_b)}} \qquad L_{sh} = \sqrt{\frac{1.5(L_c)(E_b)(\text{OD})}{1{,}000(\sigma_b)}} \qquad \text{(Eq 8-7)}$$

Where:

L_{sh} = length from direction change to the first secure hanger, ft

Where:

L_{sh} = length from direction change to the first secure hanger, m

This type of analysis usually neglects torsional stresses. Allowable bending stress is much lower than the allowable torsional stress. Therefore, bending of the pipe leg, as shown in Figure 8-3, will typically absorb pipe movement. However, the unanchored leg must have a free length equal to or greater than L_{sh}, as calculated from Eq 8-7.

8.4 SUPPORTS, ANCHORS, AND GUIDES

8.4.1 Guides

The guiding mechanism must be loose to allow free axial movement of the pipe. However, the guides must be attached rigidly to the supporting structure so that the pipe moves only in the axial direction (Figure 8-4).

All guides act as supports and must meet the minimum requirements for supports. Refer to Sec. 8.4.6 if thermal cycles are frequent.

8.4.2 Anchors

An anchor must restrain the movement of the pipe against all applied forces. Pipe anchors divide a pipe system into sections. They attach to structural material capable of withstanding the applied forces. In some cases, pumps, tanks, and other similar equipment function as anchors. However, most installations require additional anchors where pipe sizes change,

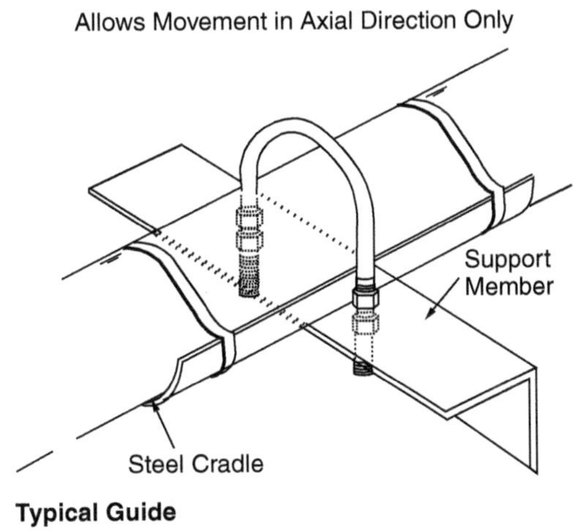

Source: Reprinted with permission from Fiberglass Pipe Handbook, Fiberglass Pipe Institute, New York

Figure 8-4 Guide support

and fiberglass pipe joins another material or a product from another manufacturer. Additional anchors usually occur at valve locations, changes in direction of piping runs, and major branch connections. Saddles and laterals are particularly sensitive to bending stresses. To minimize stresses on saddles and laterals, anchor the pipe on either side of the saddle or anchor the side run.

Figure 8-5 shows a typical anchor. Operating experience with piping systems indicates that it is a good practice to anchor long, straight runs of aboveground piping at approximately 300-ft (90-m) intervals. These anchors prevent pipe movement caused by vibration or water hammer.

One anchoring method features a clamp placed between anchor sleeves or a set of anchor sleeves and a fitting. The sleeves bonded on the pipe prevent movement in either direction. Sleeve thickness must equal or exceed the clamp thickness. To achieve this, it is often necessary to bond two sleeves on each side of the clamp. Anchor sleeves are usually one pipe diameter in length and cover 180° of circumference. Anchors act as supports and guides and must meet minimum requirements for supports.

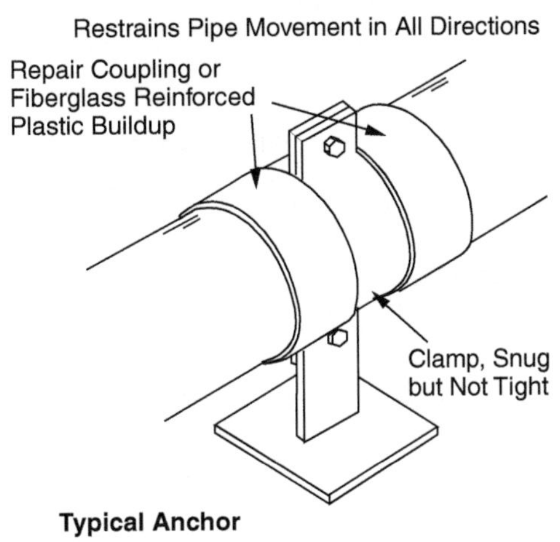

Source: Reprinted with permission from Fiberglass Pipe Handbook, Fiberglass Pipe Institute, New York

Figure 8-5 Anchor support

8.4.3 Supports

To prevent excessive pipe deflection due to the pipe and fluid weight, support horizontal pipe (see Figure 8-6) at intervals determined by one of the following methods.

8.4.3.1 Type I. Pipe analyzed as simply supported single spans (two supports per span length) with the run attached to a fitting at one end, or any other section of less than three span lengths. Beam analysis for other types of spans, such as a section adjacent to an anchor, is sometimes used to obtain a more accurate span length. However, the following equation is more conservative:

$$L_s = \sqrt[4]{\frac{(d_m)(E_b)(I)}{0.013(W)}} \qquad L_s = \sqrt[4]{\frac{(d_m)(E_b)(I) \times 10^9}{13(W)}} \qquad \text{(Eq 8-8)}$$

Where:

L_s = unsupported span, in.
d_m = allowable midpoint deflection (typical value for fiberglass pipe is 0.5 in.), in.
W = W_p + W_f, lb/in. of length
W_p = pipe weight, lb/in. of length
W_f = fluid weight, lb/in. of length
= $\rho V_p/12$
ρ = fluid density, lb/ft^3
V_p = pipe volume per foot of length, ft^3/ft

Where:

L_s = unsupported span, m
d_m = allowable midpoint deflection (typical value for fiberglass pipe is 12.5 mm), mm
W = W_p + W_f, N/m of length
W_p = pipe weight, N/m of length
W_f = fluid weight, N/m of length
= $\rho V_p/12$
ρ = fluid density, N/m^3
V_p = pipe volume per meter of length, m^3/m

When midspan deflection d_m exceeds 0.5 in. (13 mm), check with the pipe manufacturer for other considerations, such as the allowable bending stress or bearing stress. When the midspan deflection is limited to 0.5 in. (13 mm), the bending stress on the pipeline is typically less than the allowable bending stress for the pipe. For installations that result in more than 0.5 in. (13 mm) of midspan deflection, the 8:1 safety factor on bending stress has proved to be sufficient to compensate for the combination of bending stress and the longitudinal stresses resulting from internal pressure.

In fact, cyclic bending tests have shown that the stresses are not additive as expected and that the 8:1 safety factor is conservative. Cyclic bending tests consist of cyclic pressure testing of pipe bent to stress levels at or greater than the design bending stress.

For low stiffness pipe with a relatively thin wall, the local bearing pressure at supports is often significant. Supports for this application usually require 180° contact and follow a conservative design-allowable bearing pressure (45 psi [310 kPa]) compared

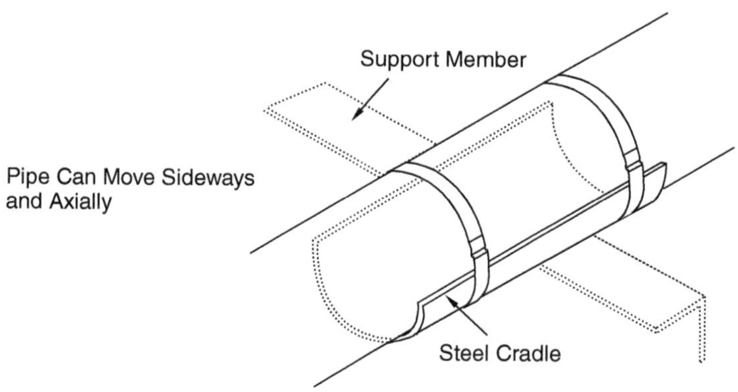

Source: Reprinted with permission from Fiberglass Pipe Handbook, Fiberglass Pipe Institute, New York

Figure 8-6 Typical support

with the typically permitted 85 psi (586 kPa) used for smaller-diameter, higher-stiffness pipe. Because pipe design differs among manufacturers, the supplier's recommendations should be followed for the product and system.

8.4.3.2 Type II. Pipe analyzed as a continuous beam—three spans—all loaded.

$$L_s = \sqrt[4]{\frac{(d_m)(E_b)(I)}{0.0069(W)}} \qquad L_s = \sqrt[4]{\frac{(d_m)(E_b)(I) \times 10^9}{6.9(W)}} \qquad \text{(Eq 8-9)}$$

8.4.3.3 Type III. Pipe analyzed as a continuous beam—four spans—all spans loaded.

$$L_s = \sqrt[4]{\frac{(d_m)(E_b)(I)}{0.0065(W)}} \qquad L_s = \sqrt[4]{\frac{(d_m)(E_b)(I) \times 10^9}{6.5(W)}} \qquad \text{(Eq 8-10)}$$

8.4.3.4 Type IV. Pipe analyzed as a beam fixed at both ends—uniformly distributed loads.

$$L_s = \sqrt[4]{\frac{(d_m)(E_b)(I)}{0.0026(W)}} \qquad L_s = \sqrt[4]{\frac{(d_m)(E_b)(I) \times 10^9}{2.6(W)}} \qquad \text{(Eq 8-11)}$$

Supports must also meet the minimum requirements for supports described in Sec. 8.4.4 through Sec. 8.4.9.

NOTE: In cases in which the wall-thickness-to-diameter ratio is low, the possibility of buckling failures at the supports is a concern. This may require the use of empirical equations and special bearing stress calculations that were determined or verified by testing.

Six basic rules control design and positioning for supports, anchors, and guides. These rules are described in the following paragraphs.

8.4.4 Rule 1: Avoid Point Loads

Use curved supports fitted to contact the bottom 120° of the pipe and that have a maximum bearing stress of 85 psi (586 kPa). Do not allow unprotected pipe to press against roller supports, flat supports like angle iron or I-beams, or U-bolts. Do not allow pipe to bear against ridges or points on support surfaces. Use metal or fiberglass sleeves to protect pipe if these conditions exist.

8.4.5 Rule 2: Meet Minimum Support Dimensions

Standard pipe supports designed for steel pipe can support fiberglass pipe if the minimum support widths provided in Table 8-1 are met. Supports failing to meet the minimum must be augmented with a protective sleeve of split fiberglass pipe or metal. In all cases, the support must be wide enough that the bearing stress does not exceed 85 psi (586 kPa).

Sleeves augmenting supports must be bonded in place using adhesives stable at the system's maximum operating temperature.

Prepare all pipe and sleeve bonding surfaces by sanding the contacting surfaces.

8.4.6 Rule 3: Protect Against External Abrasion

If vibrations or pulsations are possible, protect contacting surfaces from wear (Figure 8-7). When frequent thermal cycles, vibrations, or pulsating loadings affect the pipe, all contact points must be protected. This is typically accomplished by bonding a wear saddle made of fiberglass, steel, or one-half of a section of the same pipe to the wall.

Table 8-1 Minimum support width for 120° contact supports

Pipe Size		Minimum Support Width	
in.	*mm*	*in.*	*mm*
1	25	0.88	22.4
1.5	40	0.88	22.4
2	50	0.88	22.4
3	80	1.25	31.8
4	100	1.25	31.8
6	150	1.50	38.1
8	200	1.75	44.5
10	250	1.75	44.5
12	300	2.00	50.8
14	350	2.00	50.8
16	400	2.50	63.5

NOTE: Table is based on maximum liquid specific gravity of 1.25.

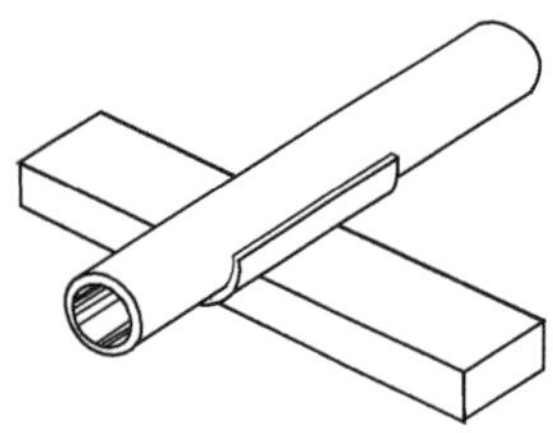

Source: Reprinted with permission from Fiberglass Pipe Handbook, Fiberglass Pipe Institute, New York, N.Y.

Figure 8-7 Fiberglass wear protection cradle

8.4.7 Rule 4: Support Heavy Equipment Independently

Valves and other heavy equipment must be supported independently in both horizontal and vertical directions (Figure 8-8).

8.4.8 Rule 5: Avoid Excessive Bending

When laying lines directly on the surface, take care to ensure that no excessive bends exist that would impose undue stress on the pipe.

8.4.9 Rule 6: Avoid Excessive Loading in Vertical Runs

Support vertical pipe runs as shown in Figure 8-9. The preferred method is to design for "pipe in compression." If the "pipe in tension" method cannot be avoided, take care to limit the tensile loadings to be less than the recommended maximum tensile rating of the pipe. Install guide collars using the same spacing intervals used for horizontal lines (Figure 8-5).

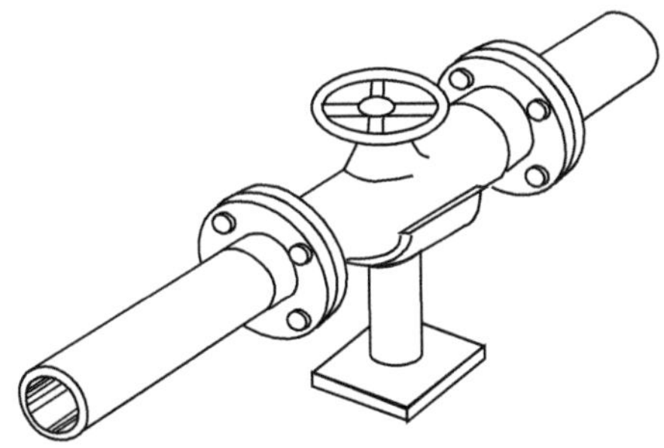

Source: Reprinted with permission from Fiberglass Pipe Handbook, Fiberglass Pipe Institute, New York, N.Y.

Figure 8-8 Steel wear protection cradle

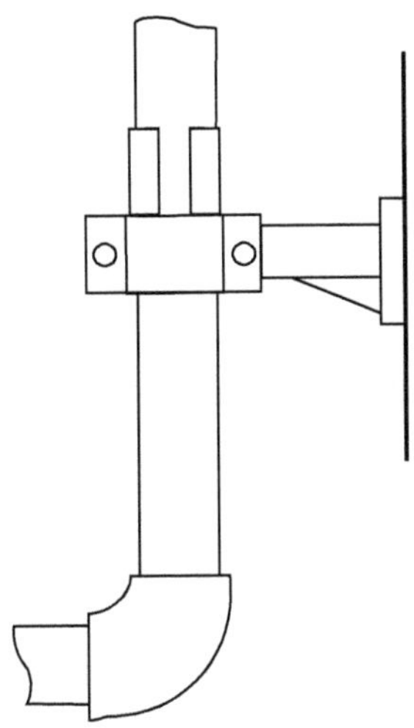

Source: Reprinted with permission from Fiberglass Pipe Handbook, Fiberglass Pipe Institute, New York, N.Y

Figure 8-9 Vertical support

8.5 BENDING

The minimum bending radius for fiberglass pipe is usually determined using a design stress that is one-eighth of the ultimate short-term bending stress. Certain fittings, such as saddles and laterals, may be more susceptible to bending failure than other types. Consult the manufacturer for recommendations and limitations. The equation for calculating the minimum bending radius is

$$R_m = \frac{(E_b)(\text{OD})}{2(\sigma_b)} \qquad R_m = \frac{(E_b)(\text{OD})}{2(\sigma_b)} \qquad \text{(Eq 8-12)}$$

Where:

R_m = minimum allowable bending radius, ft

Where:

R_m = minimum allowable bending radius, m

Because material properties vary with temperature, the allowable minimum bending radius will also vary.

8.6 THERMAL CONDUCTIVITY

The thermal conductivity of fiberglass pipe wall is ~1 percent that of steel. However, in most heat transfer situations, the heat loss or gain for pipe is controlled by the resistance to heat flow into the surrounding media (i.e., air or soil) rather than the thermal conductivity of the pipe. This reduces the insulating effect of a relatively thin fiberglass pipe wall. For this reason, thermal insulation tables for steel pipe can be used to size the insulation for most fiberglass pipelines.

The coefficient of thermal conductivity varies for different types of fiberglass pipe. A typical value for an epoxy resin pipe is 2.5 to 3.0 Btu/(h) (ft^2) (°F)/in. (0.36–0.43 W/m-K). A typical value for polyester or vinyl ester resin pipe is 1.0– to 1.5 Btu/(h) (ft^2) (°F)/in. (0.14–0.22 W/m-K).

8.7 HEAT TRACING

Both steam tracing and electrical heating tapes are acceptable techniques for heating fiberglass pipe. When using either method, three criteria govern the maximum element temperature:

1. The average wall temperature must not exceed the temperature rating of the pipe.
2. The maximum tracing temperature must not be more than 100°F (38°C) warmer than the maximum rated temperature of the pipe.
3. The maximum recommended chemical resistance temperature of the pipe must not be exceeded at the inside wall of the pipe.

For criteria 1, the following equation is applicable:

$$A_t = (T_i + T_t)/2 \qquad\qquad A_t = (T_i + T_t)/2 \qquad\qquad \text{(Eq 8-13)}$$

Where:

A_t = average wall temperature, °F
T_i = inside wall temperature, °F
T_t = heat tracing temperature, °F

Where:

A_t = average wall temperature, °C
T_i = inside wall temperature, °C
T_t = heat tracing temperature, °C

For criteria 2, the following equation is used:

$$T_t = T_R + 100°F \qquad\qquad T_t = T_R + 56°C \qquad\qquad \text{(Eq 8-14)}$$

Where:

T_R = maximum rated temperature of pipe, °F

Where:

T_R = maximum rated temperature of pipe, °C

(NOTE: The 100°F is a Δ temperature, which must be converted to metric by 0.56°C/°F.)

The maximum tracing element temperature is the lesser of the values calculated using Eq 8-13 and Eq 8-14.

The maximum tracing element temperature using this methodology applies only to applications involving flowing, nonstagnant fluid conditions. For stagnant conditions, the maximum allowable trace element is the chemical resistance temperature of the pipe.

For criteria 3, the manufacturer's published data should be checked to determine the maximum recommended chemical resistance of the pipe for this application. This value is compared with the inside wall temperature T_i. The published value must be greater than T_i.

8.8 CHARACTERISTICS AND PROPERTIES

The characteristics and properties for fiberglass pipe are different from those typically used for metallic pipes.

8.8.1 Design Pressure or Stress

Design stresses for pipe internal pressure are based on ASTM D2992, Standard Practice for Obtaining Hydrostatic or Pressure Design Basis for "Fiberglass" (Glass-Fiber-Reinforced Thermosetting-Resin) Pipe and Fittings. The internal operating pressure for fittings is generally based on one-fourth of the ultimate short-term failure pressure as determined by ASTM D1599, Standard Test Method for Resistance to Short-Time Hydraulic Pressure of Plastic Pipe, Tubing, and Fittings.

8.8.2 Modulus of Elasticity

The modulus of elasticity for fiberglass pipe is different in the axial and the hoop directions because the pipe is an anisotropic composite material. Also, the tensile, bending, and compressive moduli may differ significantly; therefore, it is important to use the correct value. The moduli depend on the type of resin, amount of glass, and orientation of the glass filaments. Precise values for the moduli for specific conditions of loading and temperature should come from the manufacturer. Typical values are often obtained by drawing a tangent to the stress-strain curve at the point equal to one-fourth of ultimate failure load. The moduli may also vary with temperature.

8.8.3 Allowable Tensile or Compressive Loads

Typically, the allowable design stress is 25 percent of the ultimate short-term failure loads. These stress values can be used with the minimum reinforced wall thickness (area) to calculate the allowable maximum loads.

8.8.4 Bending Loads

Ultimate beam stress is determined by using a simple beam with a concentrated load applied to the center to achieve short-term failure. The allowable design stress is then established by application of at least an 8:1 factor of safety to the ultimate failure value. The 8:1 factor is selected to compensate for combined loading that occurs in pressure piping applications.

The bending modulus is determined from a test by measuring midspan deflections of a simply supported beam with a uniformly distributed load over time, usually not less than 6 weeks. Allowable bending stress and the bending modulus of elasticity may vary with temperature. Values must account for the temperature extremes expected to occur in the piping application under consideration.

8.8.5 Poisson's Ratio

Because fiberglass piping is an anisotropic material, Poisson's ratio varies depending on loading conditions. For example, Poisson's ratio in the transverse (hoop) direction resulting from the axial loading is not the same as Poisson's ratio in the axial direction resulting from transverse (hoop) loading.

8.8.6 Vacuum or External Pressure

Fiberglass pipe can convey materials under vacuum. The ability of fiberglass pipe to resist collapse pressure depends on the pipe stiffness, which is a function of the pipe size, method of manufacture, ratio of diameter to wall thickness, and the raw materials used.

The external pressure resistance of fiberglass pipe may be determined by testing in accordance with ASTM D2924, Standard Test Method for External Pressure Resistance of "Fiberglass" (Glass-Fiber-Reinforced Thermosetting-Resin) Pipe. This standard test method identifies two short-term failure pressures:

- Buckling failure pressure: the external gauge pressure at which buckling occurs.
- Compressive failure pressure: the maximum external gauge pressure that the pipe will resist without transmission of fluid through the wall.

Scaling constants are used to relate the test data to pipe sizes that are not tested.

Typically, the manufacturer's recommended value for collapse pressure is 33 to 75 percent of the pipe ultimate short-term external failure pressure (ASTM D2924). The manufacturer's recommended values should be used for design purposes.

8.8.6.1 Buckling scaling constant.

$$K = \frac{P_{ext}}{E_h} \times \left(\frac{r}{t}\right)^3 \qquad K = \frac{P_{ext} \times 10^{-6}}{E_h} \times \left(\frac{r}{t}\right)^3 \qquad \text{(Eq 8-15)}$$

Where:

K = buckling scaling constant
P_{ext} = external collapse pressure, psig
E_h = circumferential flexural modulus of elasticity, psi
r = mean radius, in.
t = minimum reinforced wall thickness, in.

Where:

K = buckling scaling constant
P_{ext} = external collapse pressure, kPa
E_h = circumferential flexural modulus of elasticity, GPa
r = mean radius, mm
t = minimum reinforced wall thickness, mm

8.8.6.2 Compressive scaling constant.

$$C = \frac{P_c(\text{OD} - t)}{2t} \qquad C = \frac{P_c(\text{OD} - t)}{2t} \qquad \text{(Eq 8-16)}$$

Where:

C = compressive scaling constant
P_c = compressive failure pressure, psi

Where:

C = compressive scaling constant
P_c = compressive failure pressure, kPa

8.9 DESIGN EXAMPLES

To assist in understanding the use of the various equations in this chapter, Table 8-2 presents several design example calculations.

Table 8-2 Design example calculations

		In.-lb Units		International System of Units	
Parameter	Symbol	Unit	Value	Unit	Value
Pipe outside diameter	OD	in.	2.375	mm	60
Pipe inside diameter	ID	in.	2.235	mm	56
Pipe reinforced wall thickness	T	in.	0.07	mm	2
Coefficient of thermal expansion	C_t	in./in./°F	1.09×10^{-5}	mm/mm/°C	1.96×10^{-5}
Axial compressive modulus at minimum temperature	E_c	psi	1.3×10^6	GPa	8.9
Axial compressive modulus at maximum temperature	E_c	psi	0.6×10^6	GPa	4.1
Axial bending modulus at minimum temperature	E_b	psi	2.2×10^6	GPa	15.2

(continued)

Table 8-2 Design example calculations *(Continued)*

		In.-lb Units		International System of Units	
Parameter	Symbol	Unit	Value	Unit	Value
Axial bending modulus at maximum temperature	E_b	psi	1.3×10^6	GPa	8.9
Axial tensile modulus	E_t	psi	1.72×10^6	GPa	11.9
Allowable bending stress	σ_b	psi	1,850	MPa	12.8
Installation temperature		°F	75	°C	24
Maximum operating temperature		°F	200	°C	93
Minimum operating temperature		°F	35	°C	2
Temperature change	T_c	°F	calc	°C	calc
Maximum temperature rating	T_r	°F	210	°C	99
Material variation factor	S_f		0.9		0.9

NOTE: The values in Table 8-2 are not direct conversions from one-unit system to the other, and in many cases, rounded values are used for ease of presentation.

In.-lb Unit Examples

***Example 1*: Determine the length change per unit length for a temperature change of 60°F.**

Use Eq 8-1.

$L_c = 12(C_t)(L)(T_c)$

$= 12(1.09 \times 10^{-5})(1)(60)$

$= 0.0078$ in./ft

***Example 2*: Determine the thermal end loads at maximum and minimum operating temperatures.**

Use Eq. 8-2.

$E_L = (C_t)(E)(A)(T_c)$

Expansion temperature change,

$T_c = 200 - 75 = 125$°F

Contraction temperature change,

$T_c = 75 - 35 = 40$°F

Cross-sectional area,

$A = \pi/4(\mathrm{OD}^2 - \mathrm{ID}^2)$

$A = 3.14/4(2.375^2 - 2.235^2)$

$A = 0.507$ in.^{2}

Expansion end load =

$(1.09 \times 10^{-5})(1.3 \times 10^6)(0.507)(125)$

use compressive mod for expansion

= 898 lb

Contraction end load =

$(1.09 \times 10^{-5})(1.72 \times 10^6)(0.507)(40)$

use tensile mod for expansion

= 380 lb

International System of Units Examples

***Example 1*: Determine the length change per unit length for a temperature change of 33°C.**

Use Eq 8-1.

$L_c = 1{,}000(C_t)(L)(T_c)$

$= 1{,}000(1.96 \times 10^{-5})(1)(33)$

$= 0.65$ mm/m

***Example 2*: Determine the thermal end loads at maximum and minimum operating temperatures.**

Use Eq. 8-2.

$E_L = (C_t)(E)(A)(T_c) \times 10^9$

Expansion temperature change,

$T_c = 93 - 24 = 69$°C

Contraction temperature change,

$T_c = 24 - 2 = 22$°C

Cross-sectional area,

$A = \pi/4(\mathrm{OD}^2 - \mathrm{ID}^2) \times 10^{-6}$

$A = 3.14/4(60^2 - 56^2) \times 10^{-6}$

$A = 0.00036$ m^2

Expansion end load =

$(1.96 \times 10^{-5})(8.9)(0.00036)(69) \times 10^9$

use compressive mod for expansion

= 4,333 N

Contraction end load =

$(1.96 \times 10^{-5})(11.9)(0.00036)(22) \times 10^9$

use tensile mod for expansion

= 1,847 N

In.-lb Unit Examples

***Example* 3:** Calculate maximum allowable guide spacing.

Use Eq 8-3.

$$L_G = \sqrt{\frac{(\pi^2)(E_b)(I)}{144(C_t)(A)(T_c)(E_c)}}$$

$$\begin{aligned} I &= \pi/64\ (\text{OD}^4 - \text{ID}^4) \\ &= 3.14/64(2.375^4 - 2.235^4) \\ &= 0.337\ \text{in.}^4 \end{aligned}$$

Calculate E_b/E_c at the minimum and maximum temperatures.

Min.: $2.2 \times 10^6/1.3 \times 10^6 = 1.69$

Max.: $1.3 \times 10^6/0.6 \times 10^6 = 2.17$

Use the lesser ratio as conservative and use the maximum temperature change of 125°F.

$$L_G = \sqrt{\frac{(3.14)^2(1.69)(0.337)}{(144)(1.09 \times 10^{-5})(0.507)(125)}}$$

$$= 7.5\ \text{ft}$$

***Example* 4:** Calculate critical buckling force.

Use Eq 8-4.

$$P_{cr} = \frac{\pi^2(E_c)(I)(S_f)}{L_G^{\ 2}}$$

$$= \frac{(3.14)^2(1.3 \times 10^6)(0.337)(0.9)}{(7.5 \times 12)^2}$$

$$= 481\ \text{lb}$$

***Example* 5:** Calculate the length of expansion loop "A" leg assuming maximum length change of 4.0 in.

Use Eq 8-5.

$$L_A = \sqrt{\frac{(K)(L_c)(E_b)(\text{OD})}{144(\sigma_b)}}$$

Assume a nonguided cantilever beam and use $K = 0.75$.

$$= \sqrt{\frac{(0.75)(4)(1.3 \times 10^6)(2.375)}{144(1{,}850)}}$$

$$= 5.9\ \text{ft}$$

If, as typically assumed, leg "B" is taken as half of leg "A," then "B" = 5.9/2 = 2.95 ft.

International System of Units Examples

***Example* 3:** Calculate maximum allowable guide spacing.

Use Eq 8-3.

$$L_G = \sqrt{\frac{(\pi^2)(E_b)(I)}{(C_t)(A)(T_c)(E_c)}}$$

$$\begin{aligned} I &= \pi/64\ (\text{OD}^4 - \text{ID}^4) \times 10^{-12} \\ &= 3.14/64(60^4 - 56^4) \\ &= 0.153 \times 10^{-6}\ \text{m}^4 \end{aligned}$$

Calculate E_b/E_c at the minimum and maximum temperatures.

Min.: $15.2/8.9 = 1.71$

Max.: $8.9/4.1 = 2.17$

Use the lesser ratio as conservative and use the maximum temperature change of 69°C.

$$L_G = \sqrt{\frac{(3.14)^2(1.71)(0.153 \times 10^{-6})}{(1.96 \times 10^{-5})(0.00036)(69)}}$$

$$= 2.3\ \text{m}$$

***Example* 4:** Calculate critical buckling force.

Use Eq 8-4.

$$P_{cr} = \frac{\pi^2(E_c)(I)(S_f) \times 10^9}{L_G^{\ 2}}$$

$$= \frac{(3.14)^2(8.9)(0.153 \times 10^{-6})(0.9) \times 10^9}{(2.3)^2}$$

$$= 2{,}284\ \text{N}$$

***Example* 5:** Calculate the length of expansion loop "A" leg assuming maximum length change of 100 mm.

Use Eq 8-5.

$$L_A = \sqrt{\frac{(K)(L_c)(E_b)(\text{OD})}{1{,}000(\sigma_b)}}$$

Assume a nonguided cantilever beam and use $K = 0.75$.

$$= \sqrt{\frac{(0.75)(100)(8.9)(60)}{1{,}000(12.8)}}$$

$$= 1.77\ \text{m}$$

If, as typically assumed, leg "B" is taken as half of leg "A," then "B" = 1.77/2 = 0.89 m.

In.-lb Unit Examples

Example 6: Calculate the length from directional change to the first secure hanger.

Use Eq 8-7.

$$L_{sh} = \sqrt{\frac{1.5(L_c)(E_b)(\text{OD})}{144(\sigma_b)}}$$

$$= \sqrt{\frac{(1.5)(4.0)(1.3\times10^6)(2.375)}{144(1,850)}}$$

$$= 8.3 \text{ ft}$$

Example 7: Calculate the allowable unsupported span length.

Use Eq 8-8.

$$L_s = \sqrt[4]{\frac{(d_m)(E_b)(I)}{0.013(W)}}$$

Assume an allowable midspan deflection (d_m) of 0.5 in., as is typical for fiberglass pipe, and a pipe weight (W_p) of 0.4 lb/ft for a 1-ft length.

$$V_p = \frac{\pi(\text{ID}^2)(L)}{4(12)^2}$$

$$= \frac{3.14(2.235^2)(1)}{4(12)^2}$$

$$= 0.0272 \text{ ft}^3/\text{ft}$$

$$W_f = (V_p)\rho$$

$$= 0.0272\ (62.4)$$

$$= 1.69 \text{ lb/ft}$$

$$W = W_f + W_p$$

$$= 1.69 + 0.4$$

$$= 2.09 \text{ lb/ft}$$

$$L_s = \sqrt[4]{\frac{(0.5)(1.3\times10^6)(0.337)}{0.013(2.09/12)}}$$

$$= 99.2 \text{ in.} = 8.26 \text{ ft}$$

Example 8: Calculate the minimum allowable bending radius.

Use Eq 8-12.

$$R_m = \frac{(E_b)(\text{OD})}{2\,(\sigma_h)}$$

$$= \frac{(2.2\times10^6)(2.375)}{24(1,850)}$$

$$= 117.7 \text{ ft}$$

International System of Units Examples

Example 6: Calculate the length from directional change to the first secure hanger.

Use Eq 8-7.

$$L_{sh} = \sqrt{\frac{1.5(L_c)(E_b)(\text{OD})}{1,000(\sigma_b)}}$$

$$= \sqrt{\frac{(1.5)(100)(8.9)(60)}{1,000(12.8)}}$$

$$= 2.5 \text{ m}$$

Example 7: Calculate the allowable unsupported span length.

Use Eq 8-8.

$$L_s = \sqrt[4]{\frac{(d_m)(E_b)(I)\times10^9}{13(W)}}$$

Assume an allowable midspan deflection (d_m) of 12.5 mm, as is typical for fiberglass pipe, and a pipe weight W_p) of 0.6 kg/m (5.9 N/m) for a 1-m length.

$$V_p = \frac{\pi(\text{ID}^2)(L)}{4(1\times10^6)}$$

$$= \frac{3.14(56^2)(1)}{4(1\times10^6)}$$

$$= 0.0025 \text{ m}^3/\text{m}$$

$$W_f = (V_p)\rho$$

$$= 0.0025\ (9,800)$$

$$= 24.5 \text{ (N/m)}$$

$$W = W_f + W_p$$

$$= 24.5 + 5.9$$

$$= 30.4 \text{ (N/m)}$$

$$L_s = \sqrt[4]{\frac{(12.5)(8.9)(0.153\times10^{-6})\times10^9}{13(30.4)}}$$

$$= 2.56 \text{ m}$$

Example 8: Calculate the minimum allowable bending radius.

Use Eq 8-12.

$$R_m = \frac{(E_b)(\text{OD})}{2\,(\sigma_h)}$$

$$= \frac{(15.2)(60)}{2(12.8)}$$

$$= 35.6 \text{ m}$$

In.-lb Unit Examples

Example 9: Determine the maximum heat tracing temperature allowed to maintain a 95°F temperature inside a pipe with a maximum temperature rating of 210°F and conveying a fluid with a maximum chemical resistance temperature of 100°F.

Criteria 1: Average wall temperature not to exceed maximum temperature rating.

Use Eq 8-13.

$$A_t = (T_i + T_t)/2$$

$$A_t = \text{average wall temperature}$$

$$= (T_i + T_t)/2 = 210$$

$$= (95 + T_t)/2 = 210$$

$$T_t = 325°\text{F}$$

Criteria 2: Maximum heat tracing temperature not to be more than 100°F warmer than the maximum rated temperature of the pipe.

Use Eq 8-14.

$$T_t = T_R + 100°\text{F}$$

$$= 210 + 100 = 310°\text{F}$$

Use the lesser of the two determinations or in this case a maximum heat tracing temperature of 310°F.

Criteria 3: The maximum recommended chemical resistance temperature of the pipe must not be exceeded at the inside pipe wall.

For this example, in flowing conditions, the inside pipe wall will not exceed the recommended maximum chemical resistance temperature of 100°F. However, if stagnant conditions could be anticipated, the heat tracing temperature should be limited to 100°F.

International System of Units Examples

Example 9: Determine the maximum heat tracing temperature allowed to maintain a 35°C temperature inside a pipe with a maximum temperature rating of 99°C and conveying a fluid with a maximum chemical resistance temperature of 38°C.

Criteria 1: Average wall temperature not to exceed maximum temperature rating.

Use Eq 8-13.

$$A_t = (T_i + T_t)/2$$

$$A_t = \text{average wall temperature}$$

$$= (T_i + T_t)/2 = 99$$

$$= (35 + T_t)/2 = 99$$

$$T_t = 163°\text{C}$$

Criteria 2: Maximum heat tracing temperature not to be more than 100°F warmer than the maximum rated temperature of the pipe. (Note: The 100°F is a Δ temperature, which must be converted to metric using 0.56°C/°F.)

Use Eq 8-14.

$$T_t = T_R + 56°\text{C}$$

$$= 9 + 56 = 155°\text{C}$$

Use the lesser of the two determinations or in this case a maximum heat tracing temperature of 155°C.

Criteria 3: The maximum recommended chemical resistance temperature of the pipe must not be exceeded at the inside pipe wall.

For this example, in flowing conditions, the inside pipe wall will not exceed the recommended maximum chemical resistance temperature of 38°C. However, if stagnant conditions could be anticipated, the heat tracing temperature should be limited to 38°C.

8.10 REFERENCES

ASTM (ASTM International). 2024. D1599. Standard Test Method for Resistance to Short-Time Hydraulic Pressure of Plastic Pipe, Tubing, and Fittings. West Conshohocken, PA: ASTM.

ASTM. 2024. D2924. Standard Test Method for External Pressure Resistance of "Fiberglass" (Glass-Fiber-Reinforced Thermosetting-Resin) Pipe. West Conshohocken, PA: ASTM.

ASTM. 2024. D2992. Standard Practice for Obtaining Hydrostatic or Pressure Design Basis for "Fiberglass" (Glass-Fiber-Reinforced Thermosetting-Resin) Pipe and Fittings. West Conshohocken, PA: ASTM.

Chapter 9

Joining Systems, Fittings, and Specials

9.1 INTRODUCTION

Several types of joining systems are available for use with fiberglass pressure pipe. Many of the systems permit joint angular deflection. Some joining systems may be designed to resist longitudinal thrust forces. Fittings and specials are available in a range of styles and configurations and are fabricated using several different manufacturing methods.

9.2 FIBERGLASS PIPE JOINING SYSTEMS CLASSIFICATION

Two general joint classifications are used: unrestrained and restrained.

9.2.1 Unrestrained Pipe Joints

Unrestrained pipe joints can withstand internal pressure but do not resist longitudinal forces. They rely on elastomeric gaskets to provide the seal. Typically, these joints can be disassembled without damage.

9.2.1.1 Fiberglass couplings or bell-and-spigot joints. These joints use an elastomeric seal located in a groove on the spigot or in the bell as the sole means to provide fluid tightness.

9.2.1.2 Mechanical coupling joints. These joints use mechanically energized elastomeric gasket seals to join two pieces of pipe. The mechanical coupling technique applies to plain-end pipe.

9.2.2 Restrained Pipe Joints

Restrained pipe joints can withstand internal pressure and resist longitudinal forces.

Joints that may later be disassembled without damage include

- coupling or bell-and-spigot with a restraining device,
- flange, and
- mechanical.

Joints that cannot be disassembled without damage or cutting apart include

- butt-and-wrap,
- wrapped bell-and-spigot, and
- bonded bell-and-spigot.

9.3 GASKET REQUIREMENTS

Gaskets used with fiberglass pipe joining systems should conform to the requirements of ASTM F477, Standard Specification for Elastomeric Seals (Gaskets) for Joining Plastic Pipe. The gasket material composition must be selected to be compatible with the intended environment.

9.4 JOINING SYSTEMS

In this section, many of the joining systems available with fiberglass pressure pipe are described; however, the details of every type of joining system available are not included. Versatility of manufacture permits differences in configuration and geometry while meeting performance requirements. Users should contact the pipe manufacturer to obtain specific details on joints and joint performance.

9.4.1 Adhesive-Bonded Joints

Three types of adhesive-bonded joints are available:

- a joint using a tapered bell and a tapered spigot (Figure 9-1),
- a straight bell and straight spigot joint (Figure 9-2), and
- a joint using a tapered bell and a straight spigot (Figure 9-3).

Adhesive-bonded joints are generally available for pipe up through 16-in. (400-mm) diameter.

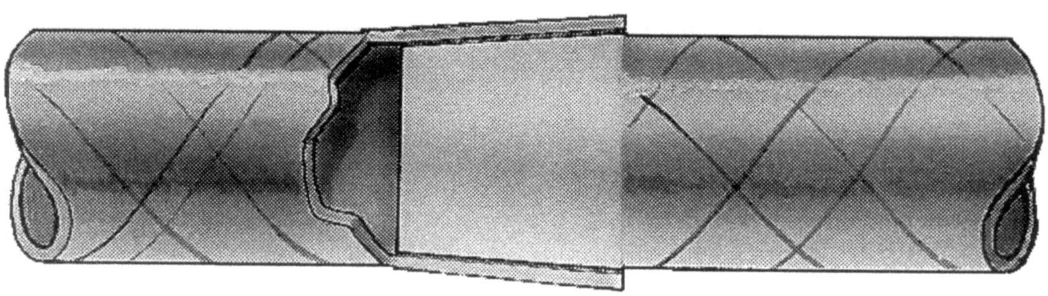

Source: Smith Fiberglass Products Inc., Little Rock, Ark.

Figure 9-1 Tapered bell and spigot joint

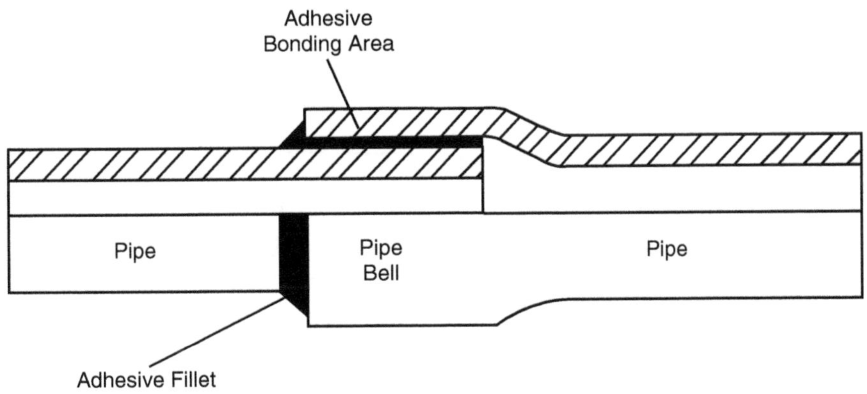

Figure 9-2 Straight bell and straight spigot joint

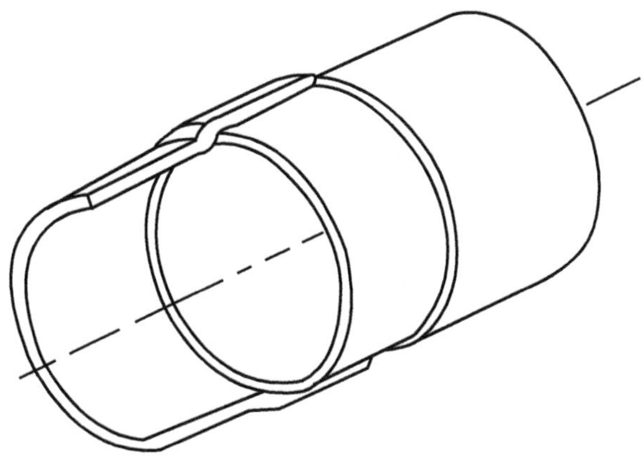

Figure 9-3 Tapered bell and straight spigot joint

9.4.2 Reinforced-Overlay Joints

The butt-and-wrap joint typically consists of two squared pipe ends that have been prepared for joining by roughening the outside surface in the joint area. The pipes are then abutted end to end, aligned on the same center line, and the joint is overwrapped with layers of resin-impregnated glass fiber materials. Each layer becomes increasingly wider to provide a buildup that accommodates internal pressure and longitudinal forces. Basic joint construction is shown in Figure 9-4, with the finished joint illustrated in Figure 9-5. A variation of this joint is illustrated in Figure 9-6, in which the pipe ends are tapered. Bell and spigot joints are sometimes overlaid, as shown in Figure 9-7. In this system, the bell aids in alignment during the overlay operation. Internal overlays are also used to improve joint performance but are generally only possible on larger-diameter pipe that allows the installer to work inside the pipe during installation.

9.4.3 Gasket-Sealed Joints

9.4.3.1 Bell and spigot. Figures 9-8 and 9-9 illustrate a bell and spigot gasketed joint using a single-gasket design. Figures 9-10 and 9-11 illustrate a bell and spigot gasketed joint using double-gasket design. The double-gasket design is generally only used with larger-diameter pipe. By inserting ports in the spigot between the two gaskets, a test of the sealing integrity of the gaskets can be conducted immediately after assembly using hydrostatic or pneumatic pressure.

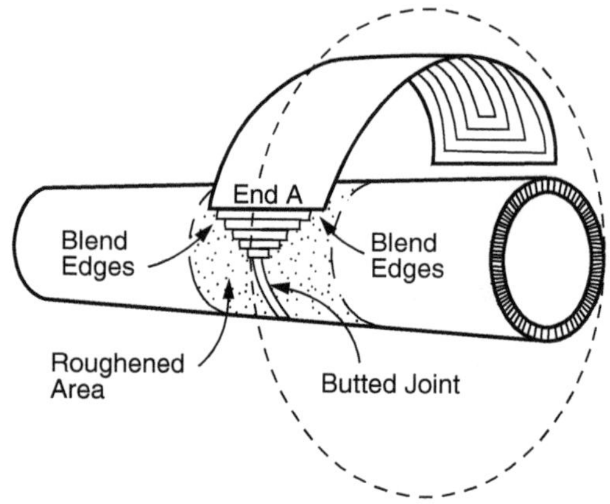

Source: Reprinted with permission from Fiberglass Pipe Handbook, Fiberglass Pipe Institute, New York

Figure 9-4 Overlay joint construction

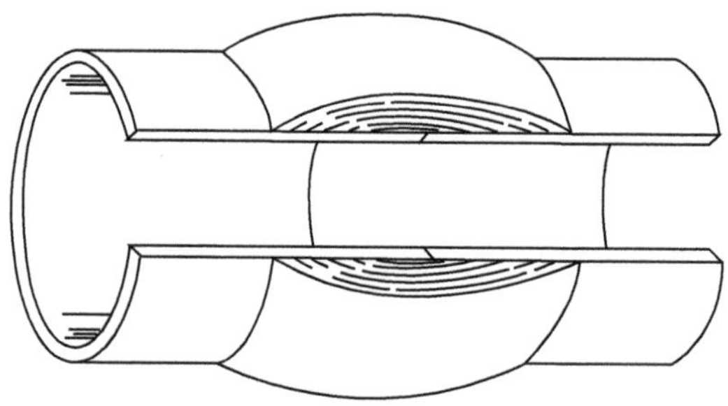

Source: Smith Fiberglass Products Inc., Little Rock, Ark.

Figure 9-5 Overlay joint

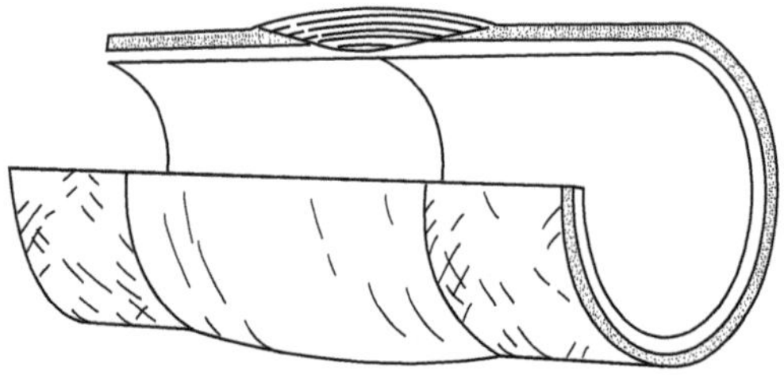

Source: Reprinted with permission from Fiberglass Pipe Handbook, Fiberglass Pipe Institute, New York

Figure 9-6 Tapered ends overlay joint

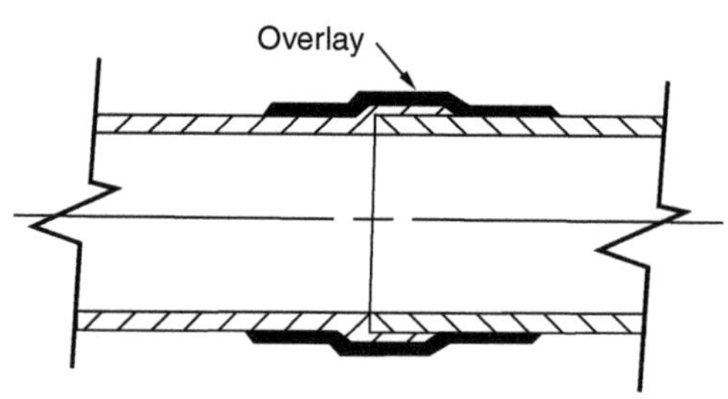

Source: Reprinted with permission from Fiberglass Pipe Handbook, Fiberglass Pipe Institute, New York

Figure 9-7 Bell and spigot overlay joint

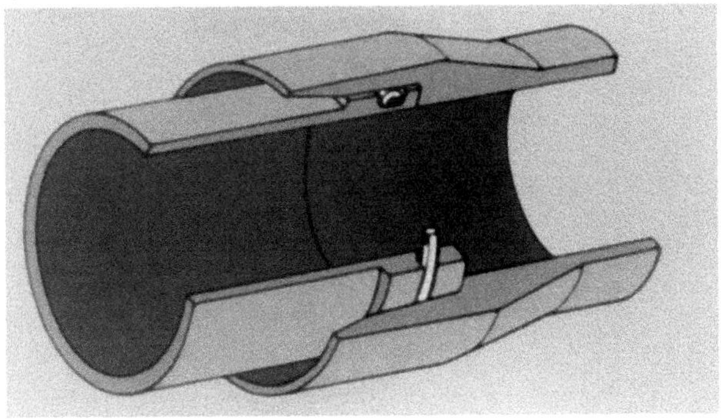

Source: Hobas Pipe USA Inc., Houston, Tex.

Figure 9-8 Single-gasket bell and spigot joint

Source: Price Brothers Company, Dayton, Ohio

Figure 9-9 Single-gasket spigot

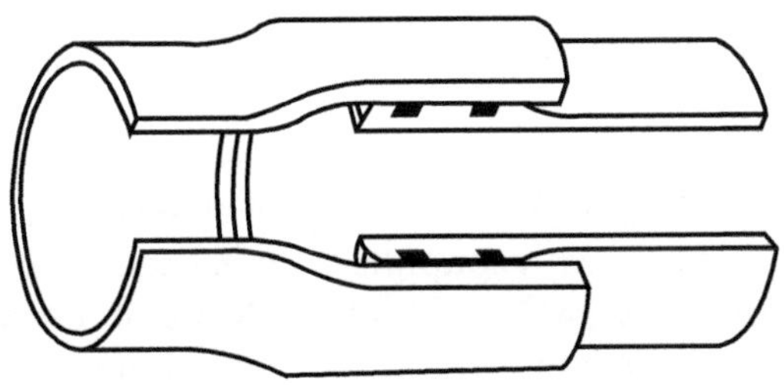

Source: Smith Fiberglass Products Inc., Little Rock, Ark.

Figure 9-10 Double-gasket bell and spigot joint

Source: Price Brothers Company, Dayton, Ohio

Figure 9-11 Double-gasket spigot

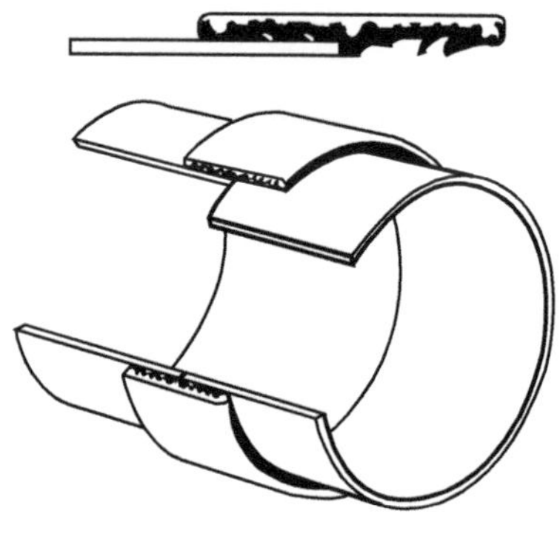

Source: Hobas Pipe USA Inc., Houston, Tex.

Figure 9-12 Gasketed coupling joint

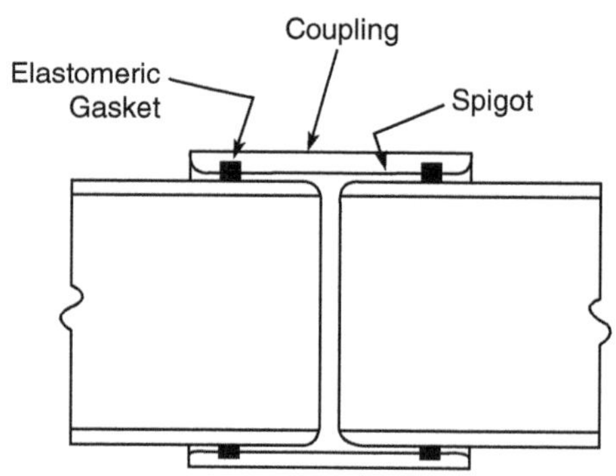

Source: Flowtite Technology, Sandefjord, Norway

Figure 9-13 Gasketed coupling joint, cross-section

9.4.3.2 Coupling. Figures 9-12 and 9-13 show two styles of gasketed coupling joints. The joint in Figure 9-12 uses a gasket mechanically bonded or molded in the coupling. Figure 9-13 shows a coupling with gaskets retained in grooves.

9.4.3.3 Restrained gasketed joints. None of the gasketed joints shown in Figures 9-8 through 9-13 provide longitudinal restraint, although they can be modified in a variety of ways to do so. Figure 9-14 illustrates a bell and spigot joint with a gasket and restraining elements. The restraining element is a mechanically loaded locking ring designed to expand and allow the spigot to enter the bell and then contract to lock on a shoulder of the spigot outside diameter. Figure 9-15 illustrates a coupling joint with a pair of gaskets and restraining elements. The shape and the material used for the restraining element can vary. Both metallic and shear-resistant plastic materials are used for this device. Figure 9-16 illustrates a bell and spigot joint with a gasket and a threaded connection joint restraining element. An advantage of many of the restrained joints is that they can be disassembled for removal or repair.

9.4.4 Mechanical Joints

Numerous mechanical joints are available for use with fiberglass pipe, including flanges, threaded joints, and commercially available proprietary joints. Pressure-rated flanges are common in the installation of all sizes of fiberglass pressure pipe. Fiberglass flanges

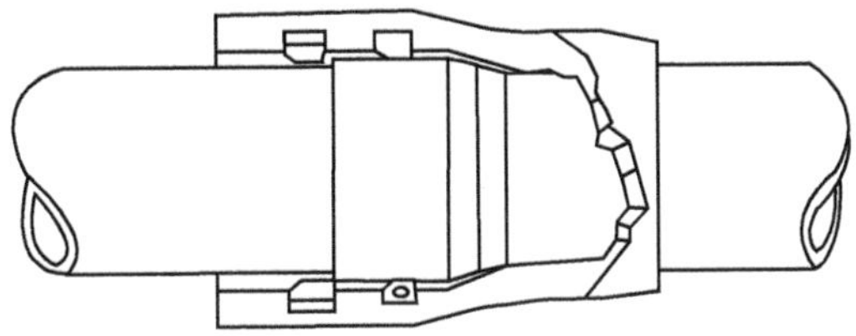

Source: Smith Fiberglass Products Inc., Little Rock, Ark.

Figure 9-14 Restrained gasketed bell and spigot joint

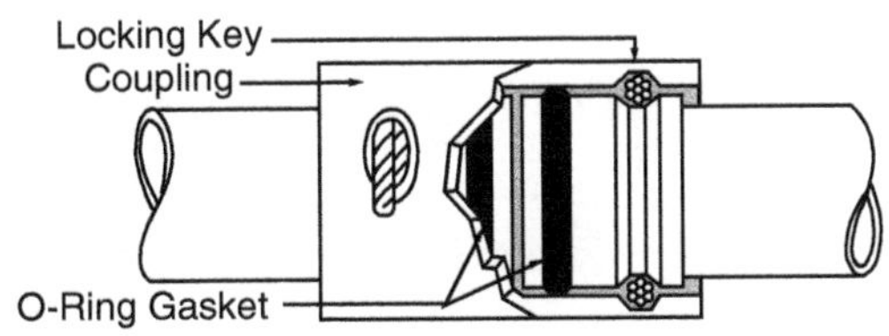

Source: Smith Fiberglass Products Inc., Little Rock, Ark.

Figure 9-15 Restrained gasketed coupling joint

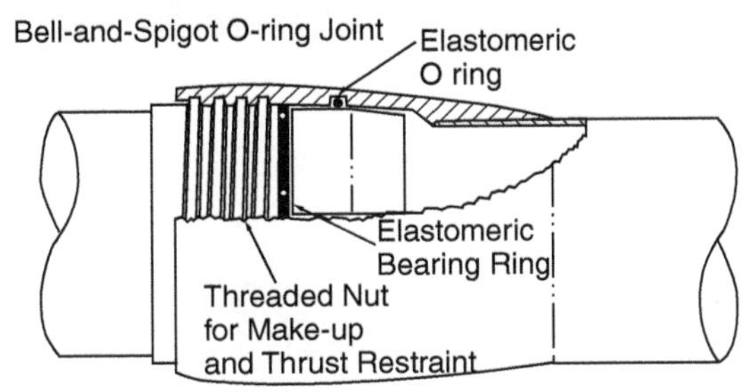

Figure 9-16 Restrained gasketed threaded bell and spigot O-ring joint

have bolting dimensions consistent with standard American National Standards Institute/American Society of Mechanical Engineers (ANSI/ASME) pressure classes of bolted flanges. Fiberglass flanges are produced by hand lay-up, filament winding, and compression molding.

Project conditions often dictate mating a fiberglass flange with a metallic flange on a pump, valve, or metallic pipe. Figure 9-17 depicts a fiberglass flange to fiberglass flange joint and a fiberglass flange to steel flange joint. Figure 9-18 shows the joining of fiberglass flanges to steel flanges to complete a valve connection.

Gaskets used with fiberglass flanges may be flat faced or O-rings contained in a groove in the flange face (see Figure 9-19). The use of O-ring seals has been found to be very effective, particularly for large diameters, because positive seal is obtained without excessive bolt torque.

Figure 9-20 shows one common mechanically coupled joint where the seal is accomplished on the outside surface of the pipe. This type of joint does not accommodate longitudinal forces. Care must be taken to not overtorque the flange joint or mechanical joint because excessive torque can damage some fiberglass pipe. A torque wrench shall be used while tightening in accordance with the manufacturer's torque limits.

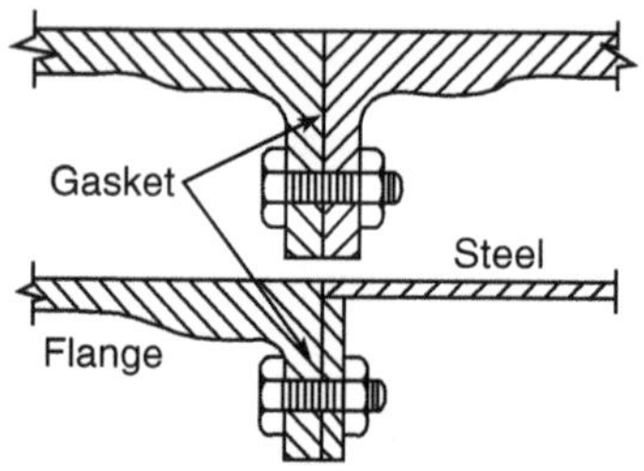

Source: Old Hope Corguard Inc. (former subsidiary of Price Brothers Company, Dayton, Ohio)

Figure 9-17 Fiberglass flange to fiberglass and steel flange joint

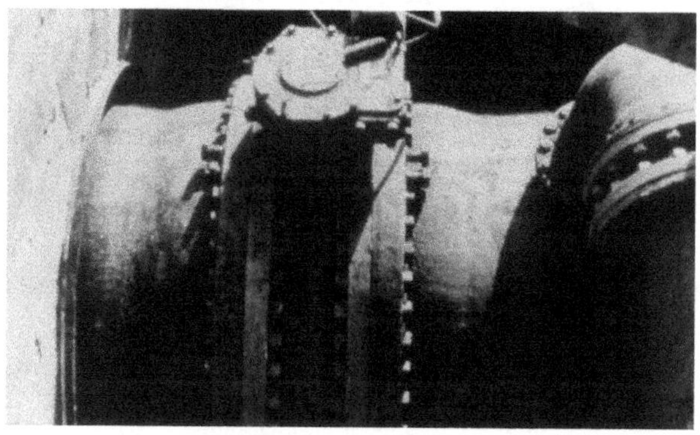

Source: Price Brothers Company, Dayton, Ohio

Figure 9-18 Fiberglass flanges to flanged steel valve connection

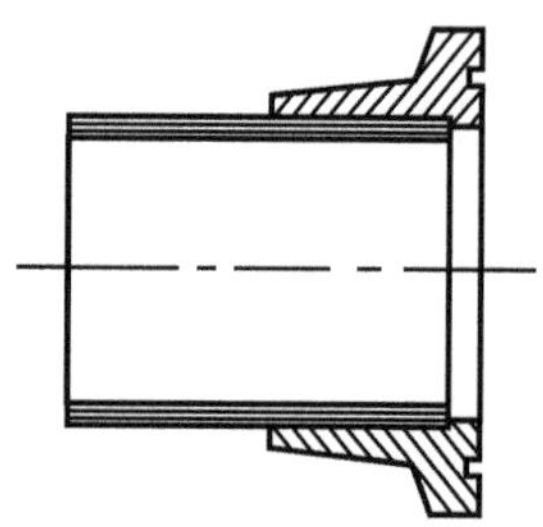

Source: Reprinted with permission from Fiberglass Pipe Handbook, Fiberglass Pipe Institute, New York

Figure 9-19 Fiberglass flange with grooved face for O-ring seal

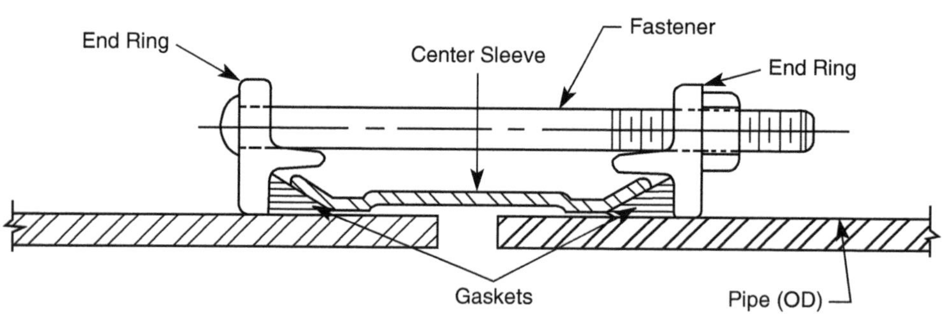

Figure 9-20 Mechanical coupling joint

9.5 ASSEMBLY OF BONDED, THREADED, AND FLANGED JOINTS

Bonded, threaded, and flanged fiberglass pipe joints require the use of techniques and equipment that may be considerably different than those used with other piping materials. Although the pipe manufacturer's instructions must always be followed, a brief general overview is given in the following sections.

9.5.1 Layout and Preparation

Installation crew members must be familiar with the installation procedures provided by the manufacturer.

Inspection of the pipe and fittings for damage that may have occurred during handling is important. Proper storage and handling procedures are discussed in Chapter 10 and are provided by the manufacturers.

The crew size requirement varies from one type of installation to another. A typical crew for 2- to 4-in. (50- to 100-mm) diameter pipe installations is two or three members, whereas installations involving large diameters can require crews of four or more members.

9.5.2 Tool and Equipment Requirements

Tool and equipment requirements vary with pipe size and type of joint; however, the following are general guidelines.

For cold weather installations, heating devices such as electric heating collars, heated portable buildings (plastic huts), hot air blowers, etc., are necessary to ensure proper installation of bonded joints.

Machining equipment such as tapering tools, disk sanders, etc., are required for end preparation on bonded joint systems. Specialized machines such as tapering tools are often available from the manufacturer. Disk grinders, belt sanders, and other, more common equipment are generally supplied by the installer.

Pipe-cutting equipment usually consists of fine-tooth saws and/or saws with carbide grit abrasive blades. Saw blades and hole saws that are typically used for wood are not suitable; however, blades used for masonry and/or tiles are usually abrasive-type blades that will be suitable for fiberglass pipe.

Note: Cutting and/or grinding operations can generate dust or cutting chips that are irritating to the skin, upper respiratory tract, and eyes. Because these materials are irritating, good ventilation for the installation crew should be used to prevent overexposure. A nuisance dust breathing filter should be used when working in areas where wind and dust are present. Tool operators should wear heavy cotton clothing, including long-sleeve shirts, that protect the skin from dust. Eye protection is often required for tool operators. Contact the local regulatory agency or Occupational Safety and Health Administration office for specific requirements on the use of respirators, protective clothing, and any additional safeguards.

Pipe chain vises and pipe stands are designed for metal piping. Therefore, it is necessary to provide protective pads such as rubber cushions to protect the pipe from point loading and/or impact damage. Protective pads are sometimes required when using come-a-longs or other tools that can create bearing and/or point loading damage.

For threaded joints, special wrenches and/or strap wrenches are recommended by most manufacturers. Caution: Improper use of strap wrenches can cause point loading.

Some tools can be used with a power drive, such as a Rigid 700 or a Rigid 300. The contractor may have to obtain a different adapter for the power drives. For example, threaded adapters used by many contractors are not used for fiberglass pipe. A typical adapter consists of a 1-in. (25-mm) drive socket that fits a $^{15}/_{16}$-in. (24-mm) square drive.

Miscellaneous equipment, such as a wraparound, felt-tip marking pens, hammers (metal and rubber), and adjustable pipe stands, are also required for installation.

9.5.3 Bonded Joint Assembly

Because many different types of joints are available, detailed assembly instructions are beyond the scope of this manual. It is essential that manufacturer's instructions be obtained for each type of joint that is being installed. The following are general guidelines.

Clean bonding surfaces are required for proper adhesion of adhesives and/or resins. In some cases, a cleaning operation such as washing and using cleaning solvent is recommended. In all cases, avoid contamination that will leave dirt, oil, grease, fingerprints, etc., on surfaces that require adhesive or resin applications. Thoroughly mix the adhesive or resin and follow safety precautions that are included with the materials. In most cases, the adhesive materials are preweighed, and it is not possible to "split a kit."

Shelf (storage) life and working (pot) life will vary from one type of resin to another. If the mixture is setting up too fast or not at all, consult the manufacturer to determine the best storage conditions, shelf life, and typical working life.

End preparation varies for the different joints. However, a clean, machined surface is generally required for application of adhesive or resin. The machining operation may involve sanding or grinding with special tools. For general sanding operations, a coarse grit (i.e., 24 or less) sandpaper is better than a fine grit.

Application of adhesives and/or resins normally requires a "wetting" process (i.e., the materials should be applied in a manner that increases the penetration and bonding of the resins to the substrate): for example, using pressure on a paintbrush to apply resin to a machined surface.

Cure times vary, and not all mixtures are properly cured when they have set up (or are hard to the touch). The proper mixing and curing procedures from the manufacturer must be followed to ensure maximum physical strength and proper chemical resistance for the system. Caution: If a mixture becomes warm and starts to cure in the container, discard it immediately. Do not use this material to assemble a joint.

In some cases, it is necessary to apply heat to speed up or ensure completion of the curing process. Caution: Allow a heated joint to cool until it is comfortable to the touch before any stress is applied to the joint. Any stress on the pipe caused by bending or sagging should be relieved before heat cure.

9.5.4 Threaded Joints

Connecting to other systems is typically accomplished with mechanical connections, threaded adapters (National Pipe Threads), reducer bushings (National Pipe Threads), grooved adapters, or flanges. Flange patterns are usually 150-lb (68-kg) or 300-lb (136-kg) bolt circle for small-diameter systems and 125-lb (57-kg) bolt circle for larger-diameter systems (greater than 24 in. [600 mm]).

Before making up threaded connections, inspect the threads. Do not use fittings with damaged threads. Inspect all metal threads. Remove any burrs and reject metal threads that have notches (grooves) that are near the end of the threads. The quality of metal threads is a concern when mating to fiberglass threads that require a low torque level. The quality of the metal threads will often have little or no effect on metal-to-metal connections because the use of additional torquing force may seal a leak. Fiberglass-to-steel connections are more likely to leak if the steel threads are in poor condition.

Unless a union is used, threaded adapters should be threaded into the other system before assembly of the fiberglass piping. Best results will be obtained using a strap wrench and a solvent-free, soft-set, nonmetallic thread lubricant. If thread sealing tapes are used, avoid improper installation of the tape, such as using thick layers of tape, to prevent damage to the fiberglass threads. In all cases, tighten the fiberglass threads as if they were brass or other soft material.

9.5.5 Flanged Joints

Most fiberglass flanges are designed for use against a flat surface; therefore, it may be necessary to use spacers or reinforcement (back-up) rings for connections to metal flanges, valves, pumps, etc. Fiberglass flanges require the use of flat washers on all bolts and nuts. In most cases, the type of gasket is specified by the manufacturer and may have a flat, O-ring, or other gasket configuration.

9.5.6 Safety Precautions

Testing with air or gas is not recommended because of the safety hazards involved. The light weight, flexibility, and elasticity of fiberglass pipe create conditions that are different from those present with steel pipe. If a catastrophic failure occurs in a fiberglass system, the system would be subject to considerable whipping and other shock-induced conditions because of the sudden release of stored energy. The recommended procedure is to conduct a hydrostatic pressure test.

9.6 FITTINGS AND SPECIALS

Fiberglass fittings and specials are available over a wide range of diameters, pressures, and configurations. Fittings and specials are made by compression molding, filament winding, cutting and mitering, and contact molding.

9.6.1 Compression Molding

Compression molding is generally used for fittings of up to 16-in. (400-mm) diameter. Figures 9-21 and 9-22 illustrate the range of configurations available for use with plain-end or flanged joints for pressure applications of less than 500 psi (3,447 kPa). In this process, a weighed glass/resin mixture is placed in a multipiece mold. The mold pieces are then held together with high pressure while the temperature is increased to cause curing. Molded fittings are cost effective for low-pressure, small-diameter applications where a large number of fittings are required.

9.6.2 Filament Winding

Filament winding can produce fittings with higher mechanical strength than is possible with molded fittings. In this process, resin-impregnated glass fibers are wound onto a

Figure 9-21 Compression-molded fittings

Figure 9-22 Flanged compression-molded fittings

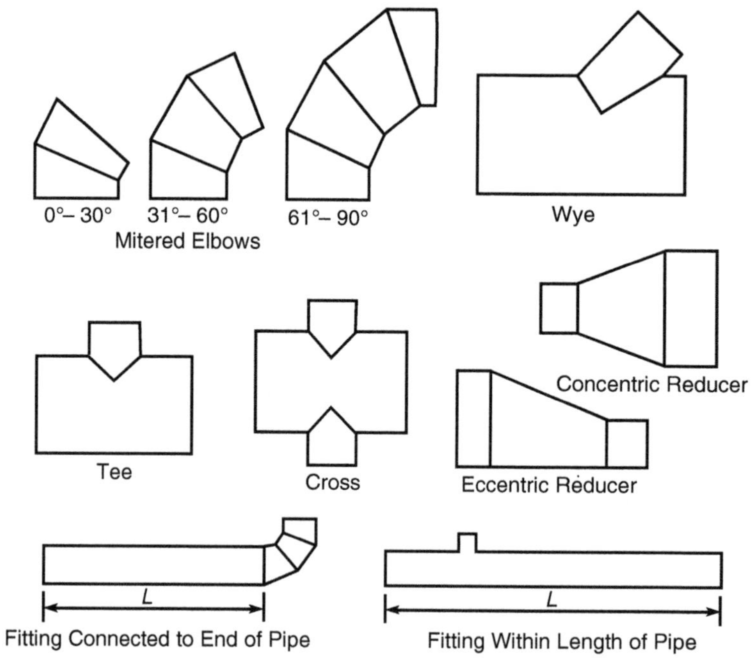

Figure 9-23 Mitered fitting configurations

fitting jig. The process may also include the use of woven rovings and/or a glass mat. After winding and curing, the fitting is removed from the jig for postproduction processing.

Cut and Miter Process. The cut and miter process is extremely versatile for making the full range of diameters, standard and special shapes, and custom-designed fittings. Figure 9-23 shows a sampling of the fittings that are routinely made from cut and mitered sections. Fabrication of the fittings and specials starts with the production of pipe that is cut and assembled into the desired configuration.

Cut and mitered fittings can also be made by cutting pipe sections to the desired form. Pieces are joined together with contact molding techniques using chopped strand and woven roving reinforcement. Surface preparation before bonding is essential to ensure good adhesion between surfaces and the contact molded laminate. The fitting should resist the same loading conditions as the pipe.

Figures 9-24 through 9-28 show a variety of fiberglass fittings and specials during fabrication and on installation sites.

Source: Price Brothers Company, Dayton, Ohio

Figure 9-24 Mitered fitting

Source: Price Brothers Company, Dayton, Ohio

Figure 9-25 Mitered fitting fabrication

Source: Price Brothers Company, Dayton, Ohio

Figure 9-26 Mitered fittings

Source: Price Brothers Company, Dayton, Ohio

Figure 9-27 Mitered fitting field fabrication

Source: Price Brothers Company, Dayton, Ohio

Figure 9-28 Fittings field assembly

9.6.3 Contact Molding

Contact molding (including spray-up) may be used to produce fittings directly.

9.7 SERVICE LINE CONNECTIONS

Service line connections are typically made using tapping saddles. Consult individual pipe manufacturers for procedures applicable to specific products.

9.8 REFERENCE

ASTM. 2021. F477. Standard Specification for Elastomeric Seals (Gaskets) for Joining Plastic Pipe. West Conshohocken, PA: ASTM International.

Chapter 10

Shipping, Handling, Storage, and Repair

10.1 INTRODUCTION

Fiberglass pipes encompass a wide range of diameters (1–156 in. [25–4,000 mm]) with an equally wide range of wall thicknesses (from less than 0.1 in. to greater than 3 in. [3–80 mm]). Furthermore, the wall laminate constructions and characteristics vary sufficiently to exhibit significantly different behaviors. Because of this wide variation in design and material characteristics, the requirements for acceptable shipping, handling, and storage are also somewhat variable. Consult the manufacturer for procedures specific to its products.

Despite the many differences, similarities also exist; therefore, several procedures occur that are typical and prudent for all fiberglass pipes. These procedures and suggestions should be used in conjunction with the pipe manufacturer's instructions. The handling requirements for fiberglass pipe are similar to those for all types of pipe.

10.2 SHIPPING

Preparation for shipping should protect the pipe wall and joining ends from damage and should be acceptable to the carrier, the manufacturer, and the purchaser.

Ship pipe on flatbed trucks supported on flat timbers or cradles (see Figure 10-1). A minimum of two supports located at the pipe quarter points is typical. Timber supports should contact only the pipe wall (no joint surfaces). No bells, couplings, or any other joint surfaces should be permitted to be in contact with the trailer, supports, or other pipe. The timber supports must be of sufficient width to avoid point loading. Chock the pipes to maintain stability and separation. To ensure that vibrations during transport do not cause

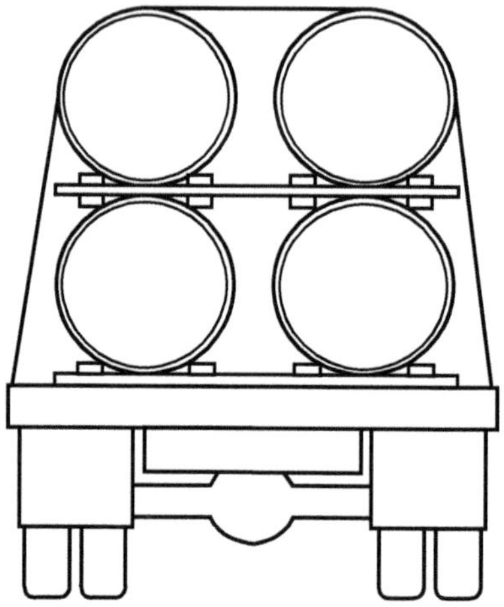

Source: Flowtite Technology, Sandefjord, Norway

Figure 10-1 Pipe shipment by truck

abrasion damage, do not allow pipes to be in contact with other pipes. Strap the pipe to the vehicle over the support points using pliable straps or rope without deforming the pipe. Bulges, flat areas, or other abrupt changes in pipe curvature are not permitted. Stack heights at the legal limits are typically acceptable.

Inspect the pipe upon receipt at the jobsite for loss or damage sustained in transit. Exterior inspection is usually sufficient; however, impact to the pipe exterior can cause interior cracking with little or no visible damage to the pipe exterior. Therefore, interior inspection at the location of exterior scrapes may be helpful when pipe size permits such an inspection. If the load has shifted or exhibits broken packaging, inspect each piece both internally and externally. Reinspect the pipe just before installation. If any imperfections or damage are found, contact the supplier for recommendations concerning repair and replacement. Note: Do not use pipe that appears damaged or defective. If in doubt, do not use. If it is necessary to transport pipes at the jobsite, use the original shipping dunnage.

10.3 HANDLING

Manufacturers' instructions regarding the use of slings, spreader bars, or other handling devices should be followed. Lift pipe sections with wide fabric straps, belts, or other pliable materials. Do not allow the straps to deform the pipe. Avoid the use of steel cables, chains, or other materials that may damage the pipe surface. If cables, chains, or forklifts are used, sufficient care, padding, or protection must be used to prevent gouging, cutting, or otherwise damaging the pipe.

Individual pipe sections can usually be lifted with a single sling (see Figure 10-2) if properly balanced, but two slings, as shown in Figure 10-3 (located at the pipe quarter points), make the pipe easier to control. Do not lift pipe with hooks or rope inserted through the pipe ends.

Because fiberglass pipe may be damaged by impact, do not drop or impact the pipe, especially the pipe ends. Pipe should never be thrown or dropped to the ground or set on sharp objects. Repair any damage before installation.

Bundles. Smaller pipe (24-in. [600-mm] diameter and less) is often unitized or bundled by the manufacturer, as shown in Figures 10-4 and 10-5. Bundles and unitized loads typically must be handled with a pair of slings (never a single sling). Do not lift a

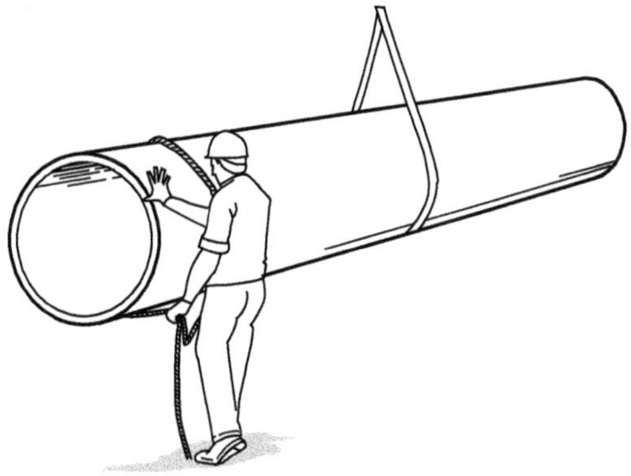

Source: Flowtite Technology, Sandefjord, Norway

Figure 10-2 Single-sling handling

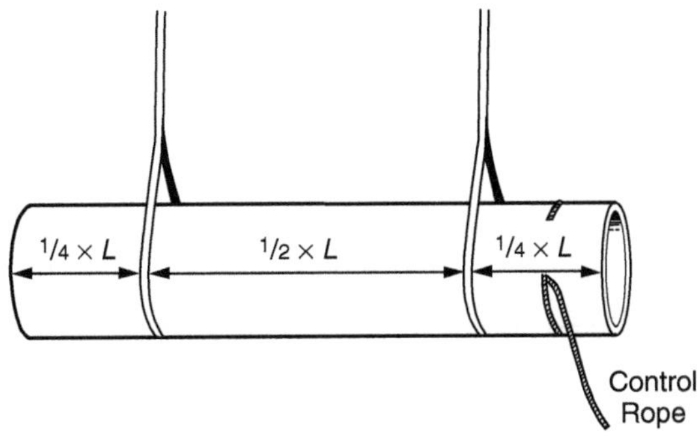

Source: Flowtite Technology, Sandefjord, Norway.

Figure 10-3 Double-sling handling

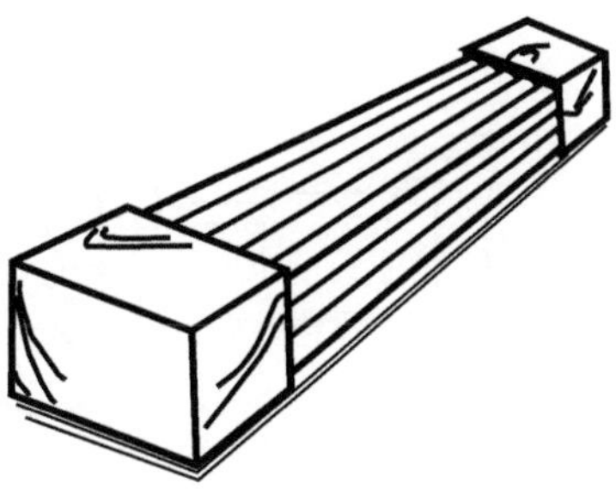

Source: Flowtite Technology, Sandefjord, Norway

Figure 10-4 Unitized small-diameter bundle

non-unitized stack of pipe as a single unit. Non-unitized stacked pipe must be unstacked and handled individually.

Nested pipe. Nesting smaller pipes inside larger pipes is acceptable. Ensure that the pipes are protected and secured properly to prevent relative motion or damage during shipment. The pipe manufacturer will provide written instructions for shipping, handling, and de-nesting of pipe. Never lift nested pipe with a single strap; always use two or more

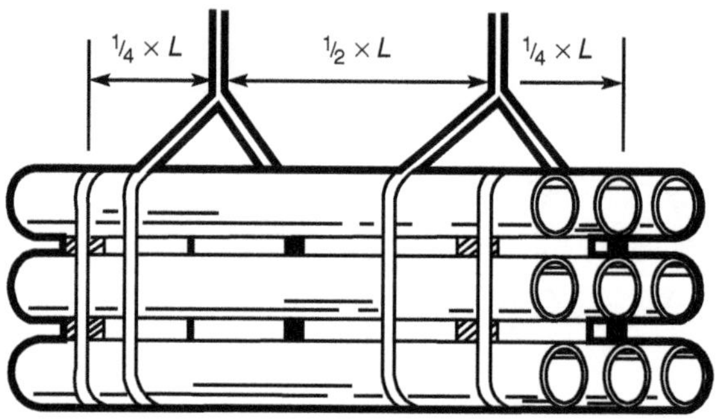

Source: Flowtite Technology, Sandefjord, Norway

Figure 10-5 Unitized load handling

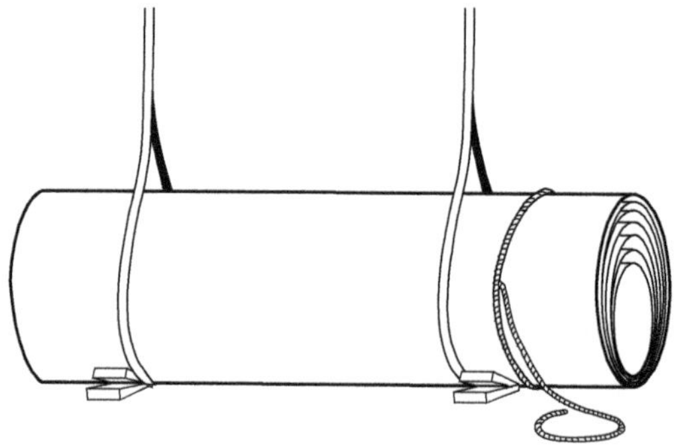

Source: Flowtite Technology, Sandefjord, Norway

Figure 10-6 Handling nested pipes

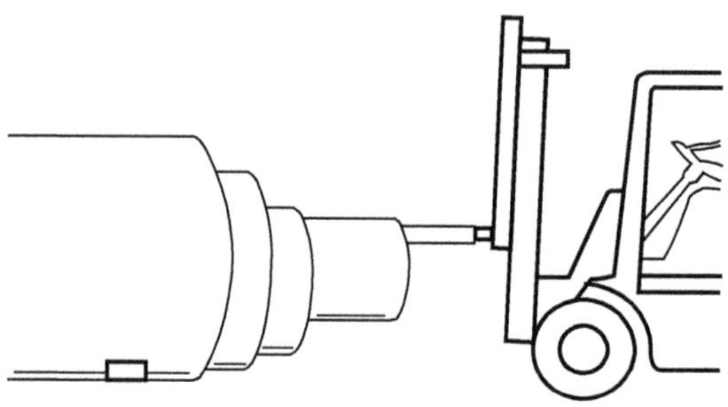

Source: Flowtite Technology, Sandefjord, Norway

Figure 10-7 De-nesting pipes

straps, as shown in Figure 10-6. Ensure that the lifting straps have the capacity to hold the bundle weight. De-nesting is typically accomplished with three or four fixed cradles that match the outside diameter of the largest pipe in the bundles. De-nest beginning with the inside pipe (smallest diameter). The standard de-nesting procedure is to insert a padded forklift boom, lift slightly to suspend the pipe, and carefully remove it without touching the other pipe (see Figure 10-7). When weight, length, and equipment limitations preclude this method, check with the manufacturer for specific recommendations for removing pipe from the bundle.

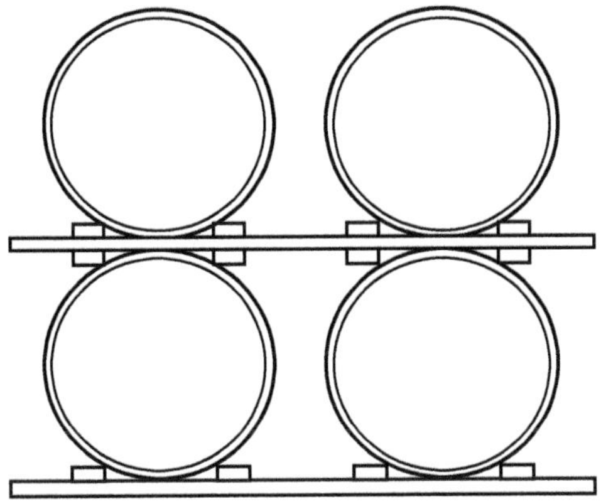

Source: Flowtite Technology, Sandefjord, Norway

Figure 10-8 Pipe stacking

10.4 STORAGE

Pipe is generally stored on flat timbers to facilitate placement and removal of lifting slings (see Figure 10-8). The support timbers should be of sufficient width to prevent point loads. Supports that are 4-in. (10-cm) wide are recommended for large-diameter pipe. Pipe should be chocked to prevent rolling in high winds. When stacking, timber supports at the pipe quarter points are best. If available, use the original shipping dunnage for storage. The maximum stack height is typically 8 ft (2.4 m). Consult the manufacturer for maximum storage deflection. Bulges, flat areas, or other abrupt changes in pipe curvature are not permitted. Nylon or hemp rope tie-downs are best. Chain tie-downs must be well padded to prevent damage to the pipe wall.

Rubber ring gaskets should be stored in the shade in the original packaging. They typically must be protected from sunlight, solvents, and petroleum-based greases and oils.

When stored directly on the ground, the pipe weight should not be supported by the bell, coupling, or any other joint surface. The pipe should rest on level ground and should not rest on rocks, boulders, or other hard debris that may cause a point load sufficient to gouge, crack, puncture, or otherwise damage the pipe wall. The pipe interior and all joining surfaces should be kept free of dirt and foreign matter.

Ultraviolet (UV) protection. Check with the pipe manufacturer regarding the necessity of UV protection when stored outside.

Nested pipe. Store nested pipe only in the original transport packaging. Do not stack nested pipe unless approved by the manufacturer. Transport pipe only in the original transport packaging.

10.5 REPAIR

Typically, damaged pipe can be repaired quickly and easily by qualified personnel at a jobsite. The repair design depends on the wall thickness, wall composition, application, and the type and extent of damage. Do not attempt to repair damaged or defective pipe without consulting the pipe manufacturer.

Scrapes and gouges on the pipe exterior that are less than 10 percent of the pipe wall thickness generally require no repair, whereas deeper scrapes generally require repair. Repair for damage to the inner liner depends on the damage depth. Scratches, scrapes,

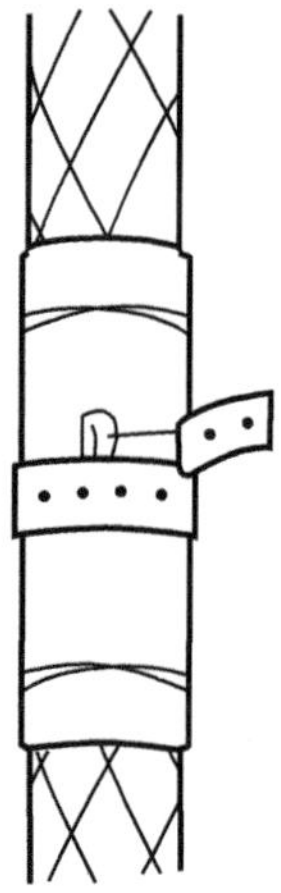

Figure 10-9 Patch

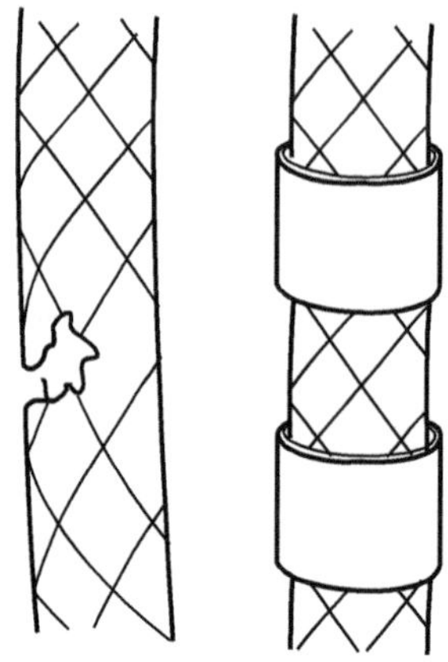

Figure 10-10 Cut out and replace

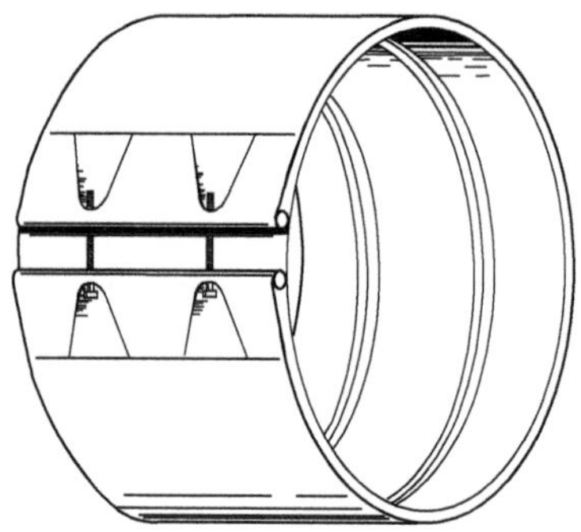

Source: Flowtite Technology, Sandefjord, Norway

Figure 10-11 Steel coupling

and abrasions that do not penetrate through the entire liner generally require no repair. Gouges through the entire liner that penetrate the interior reinforced structural glass require a mat/resin lay-up to restore the original pipe wall thickness. Structural fracture of the pipe wall is evaluated on a case-by-case basis to provide repair sufficient to restore the original pipe strength.

Damaged pipe can either be replaced or repaired. During repair, the pipeline cannot be under pressure, and the area to be repaired must be dry throughout the procedure. Repair techniques include patching small areas (Figure 10-9), cut out and replace (Figure 10-10), repair clamps, hand lay-up, and flexible steel couplings (Figure 10-11). Consult the pipe manufacturer to determine which method is appropriate.

Consult the pipe manufacturer regarding minor repairs of damaged fittings. Extensively damaged pipe and fittings must be replaced.

Hand lay-up repair. The pipe manufacturer should be contacted for job-specific lay-up instructions. Some manufacturers provide field lay-up kits that are individually prepared for the pipe diameter and pressure rating. Kits include premeasured resin, catalyst, and precut glass mat. The ambient temperature should be between 60°F and

100°F (15°C and 38°C), and the repair should be protected from the sun while curing to prevent temperature differentials. Lay-up repairs require clean, controlled conditions and skilled, trained personnel.

Flexible steel couplings can be used for joining pipe sections as well as for repairs. Steel repair couplings consist of a steel mantle with an interior rubber sleeve.

This page intentionally blank.

Glossary

Fiberglass pipe materials, processes, product standards, test methods, and installation practices and procedures may introduce some terms and terminology that are new to the unfamiliar user. This glossary provides basic definitions of fiberglass pipe terms used in this manual and by those in the fiberglass pipe industry.

accelerator – See *hardener*.

adapter – A fitting used to join two pieces of pipe, or two pipe fittings, that have different joining systems.

aggregate – Siliceous sand conforming to ASTM C33, except that the requirements for gradation do not apply.

aliphatic amine curing agent – Aliphatic amines are curing agents for epoxy resins. Aliphatic amine-cured epoxy resins cure at room temperature, a property that makes them especially suitable for use in adhesives. Some filament-wound pipes use aliphatic amine cured epoxy resins. The properties of these pipes depend on the specific amine used in manufacture.

anhydride curing agents – Anhydrides are widely used curing agents for fiberglass reinforced epoxy pipe. The properties of these pipes depend on the specific anhydride used in manufacture.

bell-and-spigot – A joining system in which two cylindrical surfaces come together to form a seal by adhesive bonding or by compression of an elastomeric gasket. The bell is the female end; the spigot is the male end.

bisphenol-A – A major ingredient used to make the most common type of epoxy resin, bisphenol-A epoxy resin. Also used as an intermediate to produce some polyester resins.

box – The female end of a threaded pipe, or fitting, connection.

buckling – See *collapse*.

burst pressure – The ultimate pressure a pipe can resist for a short term before failing. Also see *weeping*.

burst strength (hoop stress) – The circumferential stress (hoop stress) at burst pressure.

bushing – A fitting used to join two different sizes of pipe by reducing the size of the female end of the joint.

catalyst – See *hardener*.

centrifugal casting – A process used to manufacture tubular goods by applying resin and reinforcement to the inside of a mold that is rotated and heated, subsequently polymerizing the resin system. The outside diameter of the finished pipe is determined by the inside diameter (ID) of the mold tube. The ID of the finished pipe is determined by the amount of material introduced into the mold. Other materials may be introduced in the process during manufacture of the pipe.

collapse – Failure caused as the result of application of a uniform force around the total circumference of the pipe. The force may be caused by an externally applied pressure or vacuum inside the pipe. The mode of failure is usually stability related and occurs as flattening of the pipe but can be caused by compressive (shear) failure of the pipe wall.

collar – See *coupling*.

compressive force – The force that occurs when opposing loads act on a material, thereby crushing or attempting to crush it. In pipe, circumferential compressive forces may result from external pressure; longitudinal compressive forces may result from heating of an end-restrained fiberglass pipe.

coupling (collar) – A short, heavy-wall cylindrical fitting used to join two pieces of the same-sized pipe in a straight line. The coupling always has female connection ends that can be threaded or that use adhesive bonding or elastomeric seals.

creep – Deformation or strain that occurs over time when a material experiences sustained stress. Creep is expressed in inch per inch (millimeter per millimeter) per interval of time. Fiberglass pipe is subject to creep at all temperatures when subjected to stress.

cure – The hardening of a thermoset resin system by heat and/or chemical action.

cure stages – The degree to which a thermoset resin has cross-linked. In order of increasing cross-linking, the three stages include B-stage, gelled, and fully cured.

curing agent – See *hardener.*

cut and mitered fittings – Fittings manufactured by cutting, assembling, and bonding pipe sections into a desired configuration. The assembled product is then overlayed with resin-impregnated roving, mat, or glass cloth to provide required strength.

cyclic pressure rating – The pressure rating obtained as the result of performing tests in accordance with ASTM D2992, procedure A. This method rates pipe on the basis of 150 million cycles. This conservative approach results in lower pressure ratings for pipes than static testing but is useful in comparing competitive products.

design factor (factor of safety) – A number equal to or greater than 1.0 that takes into consideration the variables and degree of safety involved in a design. Test data are divided by the design factor to obtain design allowable values. It is the reciprocal of the service factor. Also called *safety factor.*

drift diameter – A measure of the effective minimum inside diameter of a pipe including ovality and longitudinal warpage over a given length of pipe.

elastic limit – See *proportional limit.*

elastic modulus (modulus of elasticity) – The "resistance" of a material to movement. The slope of the stress-strain curve within the elastic range.

epoxy resin (thermosetting) – A polymer containing two or more three-membered rings, each consisting of one oxygen and two carbon atoms. The polymer is cured by cross-linking with an amine or anhydride hardener, with or without heat, catalyst, or both.

fatigue – Permanent structural damage in a material subjected to fluctuating stress and strain.

fiberglass pipe – A tubular product containing glass-fiber reinforcements embedded in or surrounded by cured thermosetting resin. The composite structure may contain aggregate, granular, or platelet fillers; thixotropic agents; and pigments or dyes. Thermoplastic or thermosetting liners or coatings may be included.

filament winding – A process used to manufacture tubular goods by winding continuous glass-fiber roving or roving tape onto the outside of a mandrel or core pipe liner in a predetermined pattern under controlled tension. The roving may be saturated with liquid resin or pre-impregnated with partially cured resin. Subsequent polymerization of the resin system may require application of heat. The inside diameter of the finished pipe is fixed by the mandrel diameter or the inner diameter of the core pipe liner. The outside diameter of the finished pipe is determined by the amount of material that is wound on the mandrel or core pipe liner. Other materials may be introduced in the process during the manufacture of the pipe.

fillers – Extender materials added to a resin that do not affect the cure of the resin but may influence the physical and mechanical properties of the resin system and the finished product.

fitting types – The classification of fittings by the method of manufacture (i.e., molded, cut and mitered, filament wound).

gel time – The time it takes for a resin system to increase in viscosity so that flow will not occur.

glass fabric – A bidirectional reinforcing material made by weaving glass-fiber yarn.

glass fibers – A commercial grade of glass filaments with binder and sizing that are compatible with the impregnating resin.

hand lay-up –Manual methods for forming resin and fiberglass into finished pipe products. These procedures include overwrap techniques, contact molding, and hand molding. Complex shapes can be fabricated.

hardener (accelerator, catalyst, curing agent, promoter) – Any of several chemicals added to the resin, individually or in combination, that speed up the curing process or cause hardening to occur.

hoop stress – Circumferential stress. See also *burst strength.*

hydrostatic design basis (HDB) – The long-term hydrostatic hoop strength of a specific fiberglass pipe material for water service as determined by tests and detailed evaluation procedures in accordance with ASTM D2992.

integral joint – A joint configuration in which the connection is an integral part of the pipe. A length of pipe with integral joints will have one male end and one female end.

isopolyester – Unsaturated polyester based on isophthalic acid.

joining (connecting) systems – Any of a variety of methods for connecting two separate components of a piping system. Included are bell-and-spigot, threaded, coupling, and mechanical devices.

joint – A term used to describe an individual length of pipe as well as the actual joining mechanism (i.e., adhesive-bonded bell-and-spigot, threaded, gasketed bell-and-spigot, gasketed coupling, etc.).

liner – A filled or unfilled thermoplastic or thermosetting resin layer, nonreinforced or reinforced, forming the interior surface of the pipe.

matrix – The resin material used to bind reinforcements and fillers together. This resin may be epoxy or polyester and, to a large extent, dictates the temperature and chemical performance for a pipe or fitting.

minimum bending radius – The allowable deflection of the center line of a pipe before damage occurs. The radius refers to an imaginary circle of which the pipe length would be an arc.

mitered fittings – See *cut and mitered fittings.*

modulus of elasticity – See *elastic modulus.*

molded fittings – Pipe fittings formed by compressing resin, chopped fiber, and other ingredients in a mold under heat and pressure.

pin – The male end of a pipe or fitting that matches with the female end of another pipe or fitting.

pipe stiffness – A measure of the force required to deflect the diameter of a pipe ring a unit amount.

Poisson's effect (ratio) – The property of a material that causes a change in its dimensions because of a force applied perpendicular to the plane of the dimension change. Expressed as the ratio of lateral strain to load direction strain.

polyester resin (thermosetting) – An ethylenic unsaturated polymer with two or more ester groups, dissolved in a reactive diluent with vinyl unsaturation. The polymer is cured through cross-linking by means of a free radical–initiated curing mechanism, such as peroxide catalyst and heat.

Any of a large family of resins that are normally cured by cross-linking with styrene. The physical and chemical properties of polyester resins vary greatly. Some have excellent chemical and physical properties, whereas others do not. Vinyl esters are a specific type of polyester resin. Polyester resins with properties suitable for use in the manufacture of fiberglass pipe include orthophthalic, isophthalic, -bisphenol-A fumarate, and chlorendic anhydride acid polyesters. Each type of resin has particular strengths and weaknesses for a given piping application.

pressure class – The maximum sustained pressure for which the pipe is designed.

pressure rating – The maximum long-term operating pressure a manufacturer recommends for a given product. Also referred to as *design pressure*.

promoter – See *hardener*.

proportional (elastic) limit – The greatest stress a material can sustain for a short time without causing permanent deformation. It is defined by the point at which the stress-strain curve deviates from linearity. For composite materials, this point is called the *apparent elastic limit* because it is an arbitrary approximation on a nonlinear stress-strain curve. See also *stress-strain diagram*.

reducer – A pipe fitting used to join two different-sized pieces of pipe. With the same center line in both pipes, the reducer is concentric; if center lines are offset, it is eccentric.

reinforced polymer mortar pipe (RPMP) – A fiberglass pipe with aggregate.

reinforced thermosetting resin pipe (RTRP) – A fiberglass pipe without aggregate.

reinforcement – Glass fibers used to provide strength and stiffness to a composite material. The form of reinforcement plays a major role in determining the properties of a composite. The fiber diameter and the type of sizing used are also factors. Terms relating to the physical form of the reinforcement include:

chopped fiber—Continuous fibers cut into short (0.125–2.0 in. [3.2–50 mm]) lengths.

filament—A single fiber of glass (e.g., a monofilament).

mats—A fibrous material consisting of -random-oriented, chopped, or swirled filaments, loosely held together with a binder.

milled fibers—Glass fibers, ground or milled into short (0.032–0.125 in. [0.81–3.2 mm]) lengths.

roving—A collection of parallel glass strands or filaments coated with a finish or coupling agent to improve compatibility with resins, gathered without mechanical twist. Roving may be processed in a continuous or chopped form.

yarn—Glass-fiber filaments twisted together to form textile type fibers.

yield—The number of yards of material made from one pound of product.

resin – Any class of solid or pseudo solid organic materials, often of high molecular weight, with no definite melting point. In the broad sense, the term is used to designate any polymer that is a basic material for plastics.

restrained pipe joints – Pipe joints that can withstand internal pressure and resist longitudinal forces. Some types may be disassembled without damage, but other types cannot be disassembled without damage or cutting apart.

service factor – A number less than or equal to 1.0 that takes into consideration the variables and degree of safety involved in a design. The service design factor is multiplied by test values to obtain design allowables. It is the reciprocal of the design factor.

static pressure rating – The recommended constant pressure at which pipe can be operated continuously for long periods without failure. Determined by conducting tests in accordance with ASTM D2992, procedure B.

stiffness class – The nominal stiffness of a specified pipe.

strain – Dimensional change per unit of length resultant from applied force or load. Measured in inch per inch (millimeter per millimeter).

stress – The force per unit of cross-sectional area. Measured in per square inch (kilopascals).

stress-strain diagram – A graphic presentation of unit stress versus the corresponding unit strain. As the load increases, elongation or deformation of the material also increases.

support spacing – The recommended maximum distance between pipe supports to prevent excessive pipe deformation (bending).

surface layer – A filled or unfilled resin layer, nonreinforced or reinforced, applied to the exterior surface of the pipe structural wall.

surfacing mat – A thin mat of fine fibers used primarily to produce a smooth surface on a reinforced plastic. Also called *surfacing veil.*

surge allowance – That portion of the surge pressure that can be accommodated without changing pipe pressure class. The surge allowance is expected to accommodate pressure surges usually encountered in typical water distribution systems.

surge pressure – A transient pressure increase greater than working pressure, sometimes called *water hammer,* that is anticipated in a system as a result of a change in the velocity of the water, such as when valves are operated or when pumps are started or stopped.

tape – A unidirectional glass-fiber reinforcement consisting of rovings knitted or woven into ribbon form.

tensile force – A force applied to a body tending to pull the material apart.

thermal conductivity – The rate at which a material transmits heat from an area of high temperature to an area of lower temperature. Fiberglass pipe has low thermal conductivity.

thermal expansion – The increase in dimensions of a material resulting from the application of heat. Thermal expansion is positive as temperature increases and negative as temperature decreases.

thermoplastic resin – A plastic that can be repeatedly softened by heating and hardened by cooling and that, in the softened state, can be fused or shaped by flow.

thermoset resin – A polymeric resin cured by heat or chemical additives. Once cured, a thermoset resin becomes essentially infusible (cannot be remelted) and insoluble. Thermosetting resins used in pipe generally incorporate reinforcements. Typical thermosets include:

epoxies

- amine cured
- anhydride cured
- aliphatic polyanhydrides
- cycloaliphatic anhydrides
- aromatic anhydrides

novolac or epoxy novolac

unsaturated polyesters

- orthophthalic polyester
- isophthalic polyester
- bisphenol-A fumarate polyester
- chlorendic acid polyester

vinyl esters

- bisphenol-A methacrylates
- bisphenol-F methacrylates

thrust forces – Commonly used to describe the forces resulting from changes in direction of a moving column of fluid. Also used to describe the axial or longitudinal end loads at fittings, valves, etc., resulting from hydraulic pressure or thermal expansion.

torque – Used to quantify a twisting force (torsion) in pipe. Torque is measured as a force times the distance from the force to the axis of rotation. Torque is expressed in foot-pounds (ft-lb) or inch-pounds (in.-lb) (Newton meters [N-m]).

ultimate pressure – The ultimate pressure a pipe can resist for a short time before failing. This pressure is typically determined by the ASTM D1599 test. May also be referred to as *ultimate burst pressure.* When some fiberglass pipes are pressured to their ultimate pressure, the failure mode may be by leakage or weeping through the pipe wall rather than fracture of the pipe wall.

unrestrained pipe joints – Pipe joints that can withstand internal pressure but do not resist longitudinal forces. They rely on elastomeric gaskets to provide the seal. Typically, these joints can be disassembled without damage.

vinyl ester – A premium resin system with excellent corrosion resistance.

weeping – Leakage of minute amounts of fluid through the pipe wall.

working pressure – The maximum anticipated, long-term operating pressure of the water system resulting from normal system operation.

woven roving – A glass-fiber fabric reinforcing material made by the weaving of glass-fiber roving.

Index

Note: *f* indicates figure; *t* indicates table

U

V

W

AWWA Manuals

M1, *Principles of Water Rates, Fees, and Charges*, #30001
M2, *Instrumentation and Control*, #30002
M3, *Safety Management for Water Utilities*, #30003
M4, *Water Fluoridation Principles and Practices*, #30004
M5, *Water Utility Management*, #30005
M6, *Water Meters—Selection, Installation, Testing, and Maintenance*, #30006
M7, *Problem Organisms in Water: Identification and Treatment*, #30007
M9, *Concrete Pressure Pipe*, #30009
M11, *Steel Pipe—A Guide for Design and Installation*, #30011
M12, *Simplified Procedures for Water Examination*, #30012
M14, *Backflow Prevention and Cross-Connection Control: Recommended Practices*, #30014
M17, *Fire Hydrants: Installation, Field Testing, and Maintenance*, #30017
M19, *Emergency Planning for Water and Wastewater Utilities*, #30019
M20, *Water Chlorination/Chloramination Practices and Principles*, #30020
M21, *Groundwater*, #30021
M22, *Sizing Water Service Lines and Meters*, #30022
M23, *PVC Pipe—Design and Installation*, #30023
M24, *Planning for the Distribution of Reclaimed Water*, #30024
M25, *Flexible-Membrane Covers and Linings for Potable-Water Reservoirs*, #30025
M27, *External Corrosion Control for Infrastructure Sustainability*, #30027
M28, *Rehabilitation of Water Mains*, #30028
M29, *Water Utility Capital Financing*, #30029
M30, *Precoat Filtration*, #30030
M31, *Distribution System Requirements for Fire Protection*, #30031
M32, *Computer Modeling of Water Distribution Systems*, #30032
M33, *Flowmeters in Water Supply*, #30033
M36, *Water Audits and Loss Control Programs*, #30036
M37, *Operational Control of Coagulation and Filtration Processes*, #30037
M38, *Electrodialysis and Electrodialysis Reversal*, #30038
M41, *Ductile-Iron Pipe and Fittings*, #30041
M42, *Steel Water-Storage Tanks*, #30042
M44, *Distribution Valves: Selection, Installation, Field Testing, and Maintenance*, #30044
M45, *Fiberglass Pipe Design*, #30045
M46, *Reverse Osmosis and Nanofiltration*, #30046
M47, *Capital Project Delivery*, #30047
M48, *Waterborne Pathogens*, #30048
M49, *Quarter-Turn Valves: Head Loss, Torque, and Cavitation Analysis*, #30049
M50, *Water Resources Planning*, #30050
M51, *Air Valves: Air-Release, Air/Vacuum and Combination*, #30051
M52, *Water Conservation Programs—A Planning Manual*, #30052
M53, *Microfiltration and Ultrafiltration Membranes for Drinking Water*, #30053
M54, *Developing Rates for Small Systems*, #30054
M55, *PE Pipe—Design and Installation*, #30055
M56, *Nitrification Prevention and Control in Drinking Water*, #30056
M57, *Algae: Source to Treatment*, #30057
M58, *Internal Corrosion Control in Water Distribution Systems*, #30058
M60, *Drought Preparedness and Response*, #30060
M61, *Desalination of Seawater*, #30061
M62, *Membrane Applications for Water Reuse*, #30062
M63, *Aquifer Storage and Recovery*, #30063
M65, *On-Site Generation of Hypochlorite*, #30065

M66, *Cylinder and Vane Actuators and Controls—Design and Installation,* #30066

M68, *Water Quality in Distribution Systems,* #30068

M69, *Inland Desalination and Concentrate Management,* #30069

M77, *Condition Assessment of Water Mains,* #30077

M80, *Biological Drinking Water Treatment* #30080

www.ingramcontent.com/pod-product-compliance
Ingram Content Group UK Ltd.
Pitfield, Milton Keynes, MK11 3LW, UK
UKHW052012041026
15590UKWH00008B/2295

9 781647 172008